ENVIRONMENT, INCENTIVES AND THE COMMON MARKET

ENVIRONMENT & POLICY

VOLUME 4

The titles published in this series are listed at the end of this volume.

Environment, Incentives and the Common Market

Edited by

Frank J. Dietz
Herman R. J. Vollebergh

Erasmus University,
Rotterdam,
The Netherlands

and

Jan L. de Vries

Study Group Environment and Economics,
National Environmental Forum,
The Netherlands

SPRINGER-SCIENCE+BUSINESS MEDIA, B.V.

Library of Congress Cataloging-in-Publication Data

Environment, incentives, and the Common Market / edited by Frank J.
Dietz, and Herman R.J. Vollebergh, and Jan L. de Vries.
p. cm. -- (Environment & policy ; v. 4)
Includes index.
ISBN 978-94-010-4054-9 ISBN 978-94-011-0113-4 (eBook)
DOI 10.1007/978-94-011-0113-4
1. Environmental impact charges--European Union countries.
2. Environmental policy--European Union countries. I. Dietz, Frank
J., 1956- . II. Vollebergh, Herman R. J. III. Vries, Jan L. de.
IV. Series.
HJ5403.5.E58 1995
363.7'0094--dc20 95-552

ISBN 978-94-010-4054-9

Cover design:
© Paul le Blanc

Printed on acid-free paper

Foreword

A growing consensus in recent years suggests that environmental policy could be enhanced by such instruments as taxation or tradeable permits. These instruments have been seriously considered as alternative or additional policies among economists for a long time. However, only recently has the debate resulted in actual proposals and even implementation, especially with respect to taxes. Usually taxes and tradeable permits are called economic instruments. They are called 'economic' because they are supposed to use as few resources as possible to attain a given environmental goal.

With this volume the authors intend to contribute to the current debate on economic instruments. It is their conviction that economic instruments can substantially further the objective of sustainable development if they are used in environmental policies in a careful and proper way, that is, if the features of the instrument(s) concerned are carefully geared to the character of the specific environmental problem. To this end, exchange between the worlds of theory and practical policy is necessary, a condition which is not sufficiently met. This volume shows politicians, civil servants and members of the environmental movements in the European Union how to use economic instruments in specific environmental policies (or their alternative proposals). Furthermore, we intend to show how theoretical economics might be made more practical, which also makes this book useful as a supplementary text for courses in environmental economics. The pros and cons of the use of economic instruments in various policy cases are discussed at an intermediate level.

This volume, therefore, has a rather unique character. In contrast to most contributions to this field, it considers practical cases of design rather than the traditional general discussion of pros and cons of such instruments in either a theoretical or too practical context. Therefore, the book might be of great help for those who are considering the use of economic instruments in their area. Moreover, currently debated environmental issues at the EU level are dealt with, such as climate change, the air quality, waste and transport. The contributors are economists and social scientists from various Dutch universities and research institutions, all of whom share a concern for the earth's environment and a commitment to improve its quality. For this reason all the contributors are participants in the Study Group on Environment and Economics of the National Environmental Forum (LMO), a team advising the Dutch environmental movement on the economic aspects of environmental strategies.

Editing books always implies a lot of effort, not only on the part of the editors. We thank, first of all, the authors who were prepared to invest so much of their time in writing and rewriting their chapters until the editors, finally, were satisfied. Furthermore, Vicky Wightman improved our English with care and patience. Elbert Dijkgraaf and Ina Lareman were helpful in the inevitable computer work of modern typewriting and table production. Paul Havekes and Ursula Roestenburg showed us how modern desk-top publishing could be used to prepare a wonderful looking camera-ready manuscript. Finally, we hope the reader will enjoy our volume as much as we enjoyed editing it.

Rotterdam/Utrecht
April 1995

Frank Dietz
Herman Vollebergh
Jan de Vries

Contents of the Volume

1 Economic Instruments: Theory and Complications

Frank J. Dietz
Herman R.J. Vollebergh
Jan L. de Vries

Theme of the Volume

This volume deals with instrument choice and design in European environmental policy. The unfolding of environmental policies in many countries and on the international level during the last decades has given rise to a renewed interest in the effectiveness and efficiency of different instruments. Moreover, given the tendencies towards intensification, internationalization and integration of environmental policy, the European Union (EU), as a rather new institutional level to which national competencies of the Member States may or may not be transferred, can be seen as a promising opportunity to optimize the policy level at which policy goals are set and implemented.

In all the case studies of this book, attention is paid to this European dimension. In cases of transfrontier pollution (e.g. the chapter on transito traffic through Austria, and the one on VOC emissions) as well as continental and global issues (e.g. emissions of greenhouse gases) this is self-evident, but even policies curbing local environmental degradation need European coordination, because of unforeseen and unwanted interferences. The EU offers the perspective of a well-coordinated European approach to environmental problems. In fact, the contours of such a European environmental policy are already visible, for example in the *Fifth Environmental Action Programme.* In addition, the intention to complete the Internal Market could be an extra incentive to coordinate the environmental policies of individual Member States.

The extent to which these opportunities will be exploited in practice is something quite different. In fact, the optimization process itself proceeds like a 'Procession of Echternach', that is making three steps forward and two steps backward. There seem to be two fundamental reasons for this. First, the ultimate principle of the European Union is to exploit comparative advantages between Member States by abolishing all kinds of trade protection. As such these measures boost opportunities for economic growth. Second, the ongoing debate about the interpretation of the principle of subsidiarity causes fundamental problems with respect to the responsibility of the EU to initiate particular environmental policy measures. According to this principle, laid down in the new constitution of the EU, the Maastricht Treaty, policy measures should be taken at the lowest possible level, adequate for solving the problem at hand. Whereas the principle as such does not raise much controversy, its implementation in specific policy areas, such as environmental policy, leaves several observers skeptical as to what extent environmental issues receive enough attention.

In any case, this book does not discuss questions concerning the role the EU should play in the environmental standard setting process, nor does it pay attention to how these goals are based on preferences of Member States or the population in the Union at large. Instead we focus on the implementation of environmental policy and its translation into policy instruments. The regulatory or legal approach to environmental policy is common in both the Union and its Member States. This approach is usually based on direct interventions using some form of legislation. Polluters' compliance is mandatory in this system, although sanctions for noncompliance do not always exist. Several failures of this 'command-and-control approach' have been reported (Hahn, 1989; Faure et al., 1994). Without being complete, we mention the paralyzing effects on the economy, legitimization problems, growing bureaucracy, and increasing enforcement problems. Due to the inflexibility for individual agents, several inefficiencies occur. Such failures are grist to those economists who cogently argue in favor of using market-based incentives, such as environmental taxes and tradeable permits. The superiority of market-based incentives, or economic instruments as they are usually called, has become well established in economics since the work of Baumol and Oates (1971). However, this superiority is based on a number of simplifying assumptions, which are, in practice, at best only partially fulfilled. This raises the basic theme of this volume whether the textbook-like pleas for using economic instruments can be developed into practical policy proposals without sacrificing too much of their attractiveness.

This volume contains six attempts to do so. Chapters 2 through 7 each deal with specific environmental problems, varying from a mainly local character (nutrient leaching from agriculture, waste disposal) to global issues (CO_2 emissions). For many of these problems a more or less far-reaching policy already exists. However, the authors of these chapters argue that the performance of these policies could be substantially improved, that is, attaining the already stated policy goals at lower costs through a different policy approach. Moreover, this general claim would even hold if one explicitly allows for the cost of regulation itself as is argued in some of the following contributions. This conclusion should be appealing, because if environmental policy measures become more stringent, the cost of regulation tends to rise considerably.

The next section sets the stage for the European dimension in this volume by evaluating the extent to which the EU is serious about developing a European environmental policy and by describing their current initiatives. Next we summarize the main features of the economic approach to environmental problems in general. The simplifying assumptions used on this abstract level of economic theory are useful to establish important theoretical insights, but are of little help if one is apt to apply the policy recommendations originating from these theoretical accounts as concrete policy proposals. The major complications are discussed in the fourth section. Our conclusion is that a case-by-case approach is the only way to synchronize the problem (the environmental issue concerned), its solution (the policy goal chosen) and the road to the solution (the instruments suggested).

Contours of a European Environmental Policy

Environmental policy, as any policy field, is an amalgam of policy goals and instruments. It can be called an amalgam because in practice goals and instruments are not easily separated. It is received knowledge in policy analysis nowadays that rational policy design is illusionary, and practical decisions only reveal some kind of 'muddling through' (Wildavsky, 1984). Therefore it takes courage to separate goals from instruments and analyze policy *as if* it were rational. Nonetheless, this is what most economists still do and this book, written by economists, is no exception. So, we assume that it is possible to distinguish *analytically* between the choice of policy goals and the choice of instruments aimed at these goals. Before explaining the economic analysis in more detail in the next sections, we first of all explore what preferences the EU has expressed with respect to both its policy goals and its instrument choice.

Policy goals or preferences can be expressed in several ways. Explicit policy goals of a government, for instance, may be aspiration levels of quality characteristic of some particular aspect of an environmental (sub)system. One example is a particular deposition level of acid substances at some location in order to improve local conditions of forests. The Fifth Environmental Action Programme of the European Union reveals some policy goals which might be interpreted this way. Aspiration levels are usually not binding. One reason is that other sets of preferences exist revealing what governments are aiming at in different policy areas. If we look, for instance, at transport and its associated air pollutants, the government aims at lowering air pollutants while expanding road infrastructures at the same time, thus boosting road traffic. Such a policy and its effects reveal the value a government attaches to the aspiration levels in practice.

Table 1.1: Environmental aspiration levels of the European Union.

Environmental Issue	*Indicator*	*EU Policy Goals*
Climate Change (including ozone depletion)	- CO_2	- Stabilization in 2000 at 1990 level (further reduction after 2005 and 2010)
	- NH_4; N_2O	- Identifying messures 1994
	- CFC-11/12; CCl_4; Cl_3CH; Halones	- Ban before 1996
	- HCFCs	- Maximum on 5% CFC level 1990
Acidification	- SO_2	- 35% reduction in 2000 from 1985 level
	- NO_x	- Stabilization in 1994 on 1990 level; 30% reduction in 2000 from 1990 level
	- NH_3	- Depending on local problems
	- VOC	- 10% reduction in 1996 from 1990 level; 30% reduction in 1999
Soil Quality	- Cd, Hg and Pb	- 70% reduction in 1995
	- Dioxine	- 90% reduction in 2005 from 1985 level (waste incineration)
Waste	- Households and dangerous waste	- Stabilization at the amount of 300 kg/per capita - Ban on exports outside EU - Infrastructure for collecting, seperating and removing - 50% recycling on average in the EU of glass,paper and plastics - Recycling of consumer products
Noise	- dB(A)	- Maximum 85 dB(A) followed by 65 dB(A) in the future - No higher percentage of population exposed to 55dB(A) or more

Source: Dutch Council for the Environment (1992)

The first set of explicit policy goals are the aspiration levels revealed in the Fifth Environmental Action Programme. Table 1.1 summarizes the most important policy goals found in the Programme. Some of the more important environmental issues are summarized, ranging from global issues such as climate change to local issues like nutrient leaching. Each issue is characterized by a particular set of indicators, if available, and the policy aspirations revealed in the Fifth Environmental Action Programme of the EU. Several of the environmental issues such as nutrient leaching, climate change and waste will be explained in more detail in the subsequent chapters, while others like acidification, soil and noise will not. Furthermore, some issues such as VOC emissions and the effect of transport on several of the environmental issues will be discussed as such. Here, it should be noted that several environmental problems exist at the same time, with different local impacts on different scales and with a number of policy aspiration levels revealed by the EU.

Table 1.1 also shows that the European Commission is of the opinion that the state of the environment is currently far from satisfactory. Thus the discrepancy between the aspiration level and the current state of affairs reveals that the status quo is suboptimal. This provides a potential for welfare-improving measures taken by the Commission itself or by the governments of the Member States. For instance, if the current contribution to acidification by the Member States of the EU is judged too high from an environmental point of view (as is revealed by the emission reduction goals), a welfare judgement has been made about the current state of affairs by some policy representative. The question as to whether this judgement also reveals welfare judgements of the general public is quite another issue, which however does not alter the fact that at least some information might be derived from such opinions. Of course, as noted before, such a set of aspiration levels might not always be a reliable source of information. Therefore, we should also look at other sources of information about the *intention* of the Commission with respect to the relative position of these goals. Two other sources might be helpful here. First, statements about the relative position of environmental goals compared with other goals of economic policy, such as economic growth, reveal something about the current value of these goals. Second, currently existing European environmental law and its enforcement reveals how much the Commission has actually done to put those aspirations into practice.

With respect to the relative position of the aspiration levels of the Union, the recent policy document on *Growth, Competitiveness and Employment,* Delors' White Paper, reveals that the Union is rather serious about the environment compared with earlier documents. In its concluding chapter the White Paper states that "the current development model in the Union is leading to a suboptimal combination of two of its main resources, i.e. labour and nature. The model is characterized by an insufficient use of labour resources and an excessive use of natural resources, and results in a deterioration of the quality of life."

The EU now acknowledges explicitly the role of the environment as important for social welfare and considers it necessary to "analyse in which ways economic growth can be promoted in a sustainable way which contributes to higher intensity of employment and lower intensity of consumption of energy and natural resources" (CEC, 1993). This evolution has not only occurred at the central level but also in the different Member States, where environmental policy initiatives have been growing as well over the past few years. Thus a considerable body of statements makes clear that concern over the environment has grown and requires a more serious approach than before.

A similar development can be observed with respect to the environmental legislation in the EU. The new Treaty of the European Union, the Maastricht Treaty, now mentions care for the environment explicitly in its Preamble and in its most important Statements. Article 2 of this Treaty states that the EU strives for sustainable growth in compliance with the environment, which is a considerable upgrading of the argument compared to earlier versions of the Treaty (Liberatore, 1990). Environmental policy as a specific area of interest for the EU has been codified in Article 130R which justifies the integration in other areas of regulation of the EU, the idea of subsidiarity in environmental policy and responsibility for global environmental issues. Furthermore, Article 130S lays down the decision procedures for environmental policy, now giving scope to majority voting instead of unanimity for enacting policy measures. Finally, Article 130T leaves room for more stringent rules for Member States if they prefer to do so, although difficulties exist in practice with respect to the interpretation of the relative strength of this article compared with the others, such as Articles 100 and 113, which regulate the free movement of goods in the Common Market.

Although serious doubts still exist about *de facto* use of these *de jure* rights, the growing number of environmental regulations concerning technical standards, bans, phase-out schemes, permits and labelling shows that the EU is taking its environmental tasks more and more seriously. There are now over 200 Directives as the palpable result of this development. And as is clear from the table presented above, the areas of intervention increasingly comprise all sectors of production and consumption. Particularly the Directives on waste are starting to 'fuel back' into the production and consumption chain, forcing both consumers and producers to change behavior.

Nonetheless, there is growing concern about both the effectiveness and efficiency of these policies. The effectiveness of current Directives, whether enacted by the EU or the Member States themselves, is being challenged more and more. The Dutch case (with which we are most familiar) demonstrates that even measures in areas which are supposed to be on the right track, still meet with considerable enforcement problems. Moreover, economic growth and its current associated rise in pollution often cancels out the gains made by these measures. Also the issue of efficiency has become more important now that the costs of regulation are growing due to stricter regulations. Present-day environmental expenditures in nine Member States for which data are available, are estimated at 0.6-1.6% of GDP, while the projected annual growth of the 'environmental market' in these Member States varies between 4 and 13% (CEC, 1992a). These problems have made room for other instruments or policies to be implemented in the future (Weale, 1994). Economics might be helpful at this point in designing more efficient and/or effective policies.

With respect to the choice of *policy instruments*, the Fifth Environmental Action Programme of the EU states that "in order to get the prices right and to create market-based incentives for environmentally friendly economic behaviour, the use of economic and fiscal instruments will have to constitute an increasingly important part of the overall approach" (p. 67). Charges and levies, fiscal incentives, direct and indirect subsidies, auditing, liability schemes, and, as a future option, tradeable permits are seen as the most important categories. So far, the introduction of an energy/carbon tax has been the most noticeable policy proposal arising from this preference shift with regard to instrument choice.

Economic Analysis of Environmental Policy

In economic science environmental issues are traditionally described as negative externalities, assuming that such externalities prevent the environment from being used in accordance with the preferences of individuals. This concept of externality, which has its roots in Pigou's *Economics of Welfare* (1920), usually serves as both a normative principle for policy intervention and as a scheme for explaining why environmental degradation occurs. As a normative principle it suggests that environmental policy goals should be derived from individual preferences directly, and not from preferences of the government. As an explanatory scheme it shows why the social optimum might not be reached if not all preferences are adequately dealt with by the price system. We first explain this economic approach stripped of its more complicated aspects and then return to its normative content.

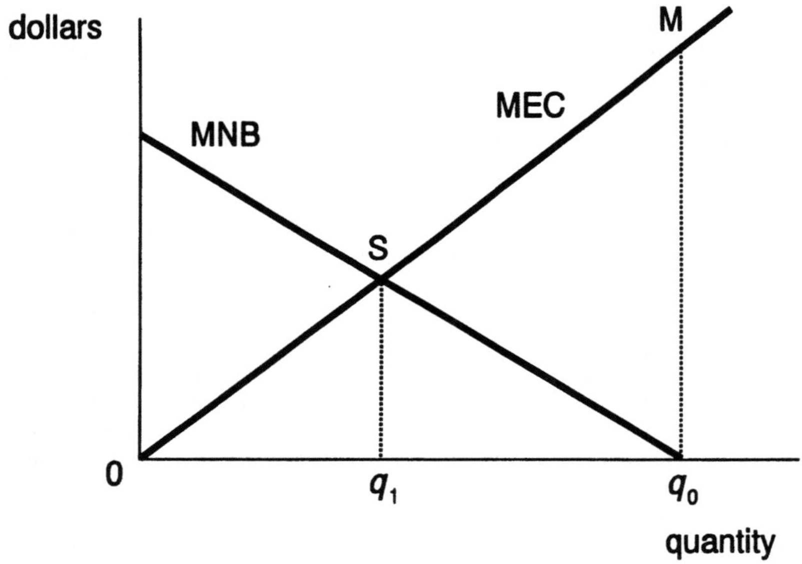

Figure 1.1: The individual and social optimum levels of production.

A negative externality has been defined as the production of a negative by-product (e.g. pollution, erosion) by one or more economic agents. This by-product is often assumed to be delivered unintentionally 'behind the back of the market'. Thus it causes a welfare loss for some victim who is not compensated for this loss, because it does not appear as a cost for the polluter(s). This basic analysis is illustrated in Figure 1.1 for the simple case that marginal environmental damage rises with production. It should be noted that this analysis is based on many simplifying assumptions, such as competitive prices on input and output markets, the flow character of the externality (pollution does not accumulate), and rationality of the economic agents such as profit maximizing producers and utility maximizing consumers (Pezzey, 1988).

Suppose that emissions of some production process generate damage by environmental degradation which is assumed to rise exponentially with the size of production (depicted as the rising linear curve for marginal environmental costs MEC). The polluters only care about the marginal net benefits of production (MNB), which is, also for reasons of simplicity, taken to be decreasing as depicted in Figure 1.1. As long as this producer maximizes profit without taking the environmental damage into account, he will increase production until the level is reached at which the marginal net benefits are zero (q_0 in Figure 1.1). Standard economic reasoning suggests this is not the socially optimal level of production, for the damage MOq_0 is not taken into account. Regarding this damage, a decrease of production would be socially desirable. The socially desirable level of production is found by weighing the social costs and benefits of various production levels. At the trajectory Oq_1, the marginal net benefits more than outweigh the marginal environmental costs, offering socially beneficial possibilities for production increase. Beyond q_1, the marginal environmental costs are higher than the marginal net benefits, requiring a decrease in production from a social point of view. Hence, incorporating adverse environmental effects in economic decisions implies a production decrease from q_0 to q_1, corresponding to the intersection S of the MNB curve and the MEC curve.

A similar reasoning applies if the environmental damage can be reduced by adapting the production process through the installment of abatement technology or by controlling the effects of emissions on the environment. In that case the MNB curve in Figure 1.1 must be interpreted as a marginal reduction cost curve (MRC curve). Along the X-axis the emission reduction percentage must be read, being 0% at q_0 and 100% at O. No marginal reduction costs exist if the producer does not abate emissions (unrestricted emissions correspond to q_0), while the abatement costs are at a maximum if the emissions are completely eliminated (corresponding with the origin of the graph O). Again the optimum corresponds with the emission reduction percentage Oq_1 where the marginal damage of the externality is equal to the marginal abatement costs.

This analysis enables economists to explain the status quo as a suboptimal state, because it is reasonable to assume that producers do not spontaneously decrease production or reduce emissions to q_1. It is, in fact, reasonable that such incentives are often absent in the case of environmental externalities. Prices for emissions, waste or diversity are often implicit or even absent, which presents at least a necessary condition for policy intervention. According to the partial approach sketched above, if adequate prices do not exist, polluters tend to produce too much of a specific externality. The social optimum S in Figure 1.1 would only be attained if society has adequate instruments which induce polluters to internalize the externality either through quantity or price-based instruments. In the next section we return to the different internalization schemes.

The externality concept is also often used as a normative principle in that the location of the optimum should be based on individual, not political preferences. It follows from the above analysis that this requires familiarity with the exact location of the social welfare optimum S. In practice, it is hard to locate this optimum, because the shape and position of the MNB curve and especially the MEC curve are hard to determine. Estimates meet with the *evaluation problem.* Problems arise especially with respect to the estimation of the damage from an increase in pollution, which is the mirror of the benefits of a decrease in environmental damage, for example, if a polluted river is cleaned up. Ideally, the Pigovian framework suggests that these benefits should be derived from individual preferences. Such preferences, and the related benefits, are very hard to discover. Some of these benefits can be derived from market prices, such as the lower costs of producing drinking water and the higher proceeds from fishing. Other benefits, however, are more difficult to estimate, simply because markets do not always exist, for instance markets for public goods like valuable ecosystems and landscapes. What is, for example, the price of a square mile of wetlands?

In the absence of markets, other evaluation methods can be used to estimate the benefits economic agents experience if particular environmental damage decreases or is avoided. In the last decades a lot of economic research has been done on alternative evaluation methods, including 'hedonic pricing methods' and 'contingent valuation methods' (CVM).[1] Although some progress has been made, these methods only *indicate* individual preferences. It is, for example, not clear whether CVM underestimates the willingness to pay for a particular environmental quality (Hoehn and Randall, 1987) or overestimates this willingness to pay (Crocker and Shogren, 1991). In addition, the crucial problem how to aggregate individual preferences into a collective statement on the value of specific natural resources is still very difficult to solve. Aggregation attempts meet with problems of the money metric measurement of utility and of interpersonal comparisons of utility. Hence, the precise shape and position of MEC curves is often hidden. The MNB curve could usually be derived from marginal reduction costs curves, which seem to be less difficult to obtain empirically. However, cost observations are not always available or reliable. To put it differently: the determination of the optimal policy intervention may imply high agency costs linked with collecting and processing the necessary information, and may even be impossible due to lack of information.

We conclude that using the externality concept based on individual preferences meets with serious difficulties. The major reason for this is that all preferences of the population at large have to be taken into account, for many environmental problems affect large numbers of people. For instance, polluters causing climate changes and victims suffering from these changes comprise many people, if not the whole global population. One can imagine how difficult a mega cost-benefit analysis of this policy issue would be. As a normative principle the Pigovian framework does not take us far in locating the optimum in practice, although much remains to be said in favor of this principle in theory. Therefore, allowing for information from outside this domain of individual preferences seems to be reasonable.

The previous section already suggested the alternative of taking preferences revealed in the policy process as a potential source of information. In theory, goals established in the political process do not necessarily imply an intrusion on voters' sovereignty (e.g. politicians as omniscient and benevolent preference brokers). In practice, however, policy goals arise as a mixture of ecological insights, ethical judgements and economic interests moulded together into goals for environmental policies. As the Public Choice literature demonstrates, it is rather unlikely that political processes produce Pareto-optimal results due to various obstacles (e.g. short memory of voters for political failures, information asymmetries, the influence of pressure groups, the power of bureaucrats). Despite these defects the environmental policy goals emerging from political processes can be taken as revealed preferences of the population at large, solving an otherwise insurmountable information problem. For that reason most of the following chapters start from this perspective.

Policy Instruments

The social optimum in the case of environmental externalities will only be attained if adequate instruments induce polluters to internalize the externality, either through quantity or price-based instruments. In theory, many kinds of instruments can be shown to solve the internalization problem as long as information and transactions are costless (Spulber, 1985). In practice, however, different internalization schemes may cause different results because information is incomplete and transactions are costly. In this section we discuss the economics of the different internalization schemes.

Perhaps the most elegant solution for the internalization problem is suggested by Coase (1960). The attractiveness of his approach from a theoretical economic point of view is that decentralized information is used which might be supposed to be in line with individual preferences. The Coasian approach also solves the problem of how to find the social optimum, because establishing the acceptable level of pollution (social optimum) is left to negotiations between polluter(s) and victim(s). Only clear and guaranteed property rights are needed for realizing agreements. However, if many parties are involved ('large number cases') and if dose-effect relations are complicated, transaction costs might become too high, preventing Coasian negotiations from taking place. In such cases a governmental agency might provide a better alternative using other internalization schemes, such as direct regulation or taxation.

In this volume, we distinguish between two broad categories of alternative internalization schemes: quantity regulation and price control. Price controlling instruments intervene indirectly in economic processes, for instance, by taxes on inputs, emissions or outputs (products), subsidies, and deposit-refund systems. Quantity regulation contains both direct interventions of governmental agencies using non-tradeable permits (command-and-control approach) and the indirect approach of issuing tradeable permits, in an amount corresponding to the environmental goal.

Figure 1.2 gives a framework for analyzing price and quantity controlling instruments. Direct intervention by the government is simply a legal rule that forces polluters to reduce their emissions, for instance by r^*%. In that case an individual polluter has no flexibility to react and simply has to comply with this rule. Another solution is depicted in the diagram, as t^* represents an environmental tax imposed per unit of pollution. The rate is equal to Sr^*, the marginal value of pollution in the social optimum. Such a tax, originally proposed by Pigou, makes it attractive for the producer to reduce his pollution to the optimum S, because beyond this point marginal reduction costs are lower than the tax rate. Dales (1968) suggests as an alternative, for either direct regulation or the Pigovian tax, the creation of a market for pollution rights. Then, r^* represents an amount of pollution rights equal to the remaining pollution, sold to the producer at price Sr^*. Also in this case the producer will reduce pollution to the optimum S, because beyond this point he has no rights at his disposal, while the cost of further reductions exceeds the price of pollution rights.

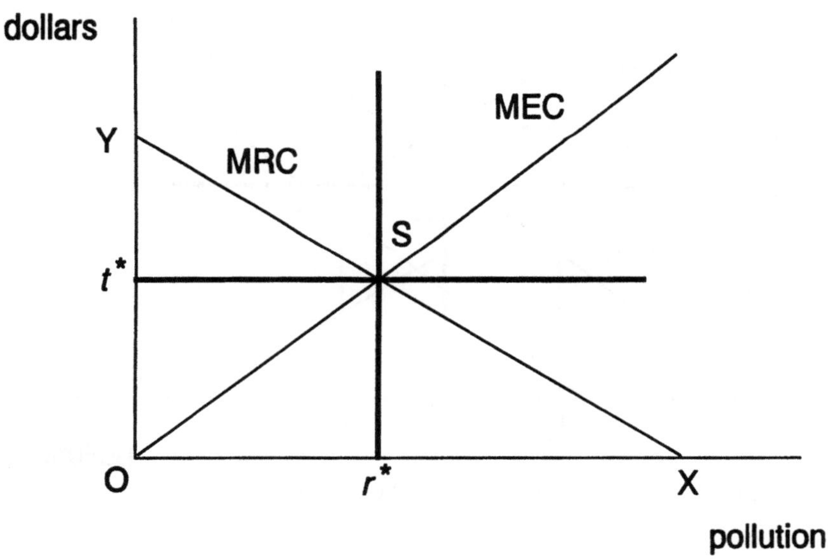

Figure 1.2: Policy interventions by price and quantity controlling instruments.

It should be noted that both the Pigovian tax at a rate of Sr^*, and the selling of pollution rights at price Sr^*, cause losses of private benefits to the producer in two ways. First, the producer bears the costs of reducing pollution to the optimum S, which are equal to the triangle r^*SX. Second, the producer has to pay taxes or to buy pollution rights for his remaining pollution, equal to Ot^*Sr^*. In this respect other price or quantity controlling instruments show different results. The most important alternatives are environmental subsidies, which reduce pollution to the optimum S without any costs for the producer, and free disposal of an amount of r^* pollution rights, implying that the producer only bears the costs of reducing the pollution r^*SX.[2] Note that legal rules only cause cost of pollution reduction, while there is no additional transfer going to the government.

A strong argument in favor of economic instruments such as taxes and tradeable permits is that they enable agents to take advantage of cost differences between them. Since Baumol and Oates' (1971) major contribution, it is well-established in economic theory that both taxes and tradeable permits are *cost-minimizing* instruments. Their theorem does not rely in any way on endogenous information about the damage of pollution or the benefits of pollution reduction.[3] They showed for the general case that any given vector of final outputs along with a set of exogenously specified standards, such as the policy preferences of the government, could be achieved at lowest cost if specific taxes are used to induce necessary adaptation of production processes. First-order conditions for cost minimization of a specified overall reduction of emissions are reproduced by cost-minimizing agents subject to the appropriate unit tax on emissions. As a result, society benefits from the fact that agents tend to optimize until the marginal cost of further reduction of emissions equals the unit tax. The marginal cost of the reduction of an externality will therefore be equalized across all activities and no opportunities exist to reduce costs by rearranging emission reduction between agents.

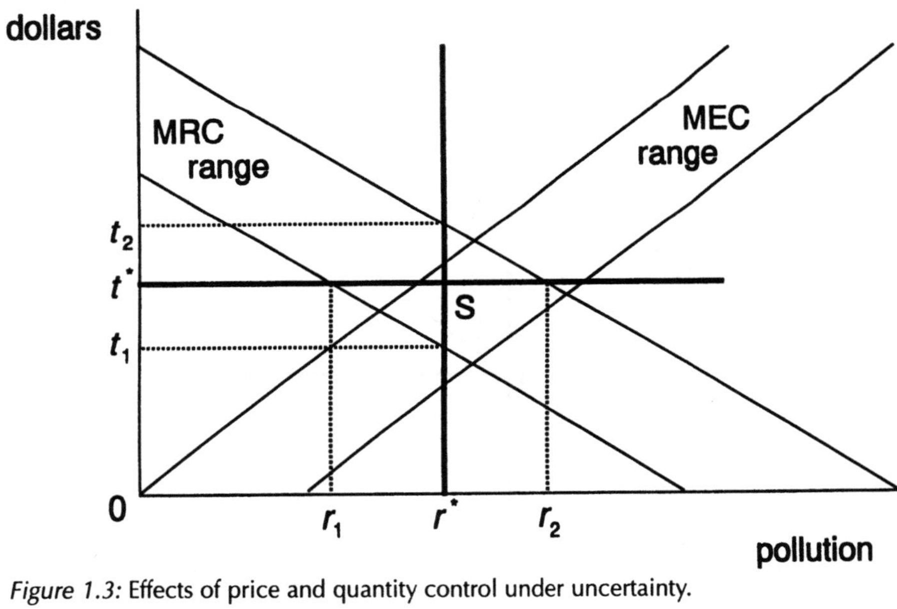

Figure 1.3: Effects of price and quantity control under uncertainty.

The same results apply if quantities rather than prices are chosen as a regulatory strategy. In the same vein, cost minimization is achieved by means of tradeable permits, which restrict the level of emissions through restricting the quantity of permits, and at the same time allows a market to develop for those permits. This market provides incentives for trade between polluters, because their marginal cost of pollution tends to differ at the outset. A market-clearing price would emerge, signaling to polluters their opportunity costs of 'wasted' emissions. Cost-minimizing behavior of agents would result in equal marginal abatement costs between sources.

The cost-efficiency argument for both taxes and tradeable permits compared to uniform direct regulation explains why these instruments are often preferred by economists. In fact, this superiority is based on a number of simplifying assumptions. Although there is no problem in finding the policy goal to be attained, for this goal is simply exogenous to the analysis, several problems remain with respect to the implementation of these theoretical devices and their associated transaction costs. As noted by a growing number of authors (e.g. Hahn, 1989; Smith, 1992; Barthold, 1994; Vollebergh, 1994), economists ignore these problems too easily and tend to be very optimistic about it. As noted in the introduction, this raises the basic theme of this volume: can the theoretical pleas for using instruments such as taxes and tradeable permits be developed into practical policy proposals without sacrificing too much of their attractiveness?

We are of the opinion that costs of implementation, administration and monitoring should also be taken into account explicitly. Like other instruments for environmental policy, taxation or tradeable permit regulation do impose such costs as well. In order to attain a reliable judgement of instruments, any regulatory scheme should be compared with other forms of regulation and their associated transaction costs. Thus, a particular policy cannot be rejected on the mere basis of its associated transaction costs as the overall principle for evaluating different forms of regulation - negotiations, direct intervention, taxes or tradeable permits - is simply which policy passes any overall welfare test, being the Pareto principle or the Kaldor-Hicks compensation test. Moreover, the precise form of the instrument also influences both the magnitude of the transaction costs and its associated benefits.

The different welfare effects of the use of various instruments is illustrated by an intervention failure due to incomplete knowledge with the regulator. Figure 1.3 shows what happens if the optimum S is not precisely targeted but is chosen somewhere between an estimated upper and lower boundary of the MEC values and similarly within the estimated range of the MRC curve. If, in this situation a price controlling instrument is applied, for example a tax with tax rate t^*, pollution will be reduced to a level somewhere between r_1 and r_2, depending on the real position of the MRC curve. If a quantity controlling instrument is used, for example tradeable pollution rights, pollution is reduced to the desired level r^*, while the price of the pollution rights will stabilize on a level between t_1 and t_2, again depending on the real position of the MRC curve. So, in the first case the uncertainty of the public agency leads to an unpredictable rate of pollution reduction, while in the second case the resulting price level is unpredictable. Moreover, in both cases the resulting cost burden for the producer, consisting of both costs of pollution reduction and either tax payments or costs of pollution rights, may be quite different from the expected burden.

This example shows one of the trade-offs that exist with respect to welfare effects of different policy instruments. To what extent specific instruments are preferable in particular situations is a question which is impossible to answer in general. As noted by Newberry (1980) economists cannot rely on theory in such situations, since they do not have an overall institutional theory providing necessary information on the cost and benefits of regulation. Here one has to rely on empirical observations. This is also the leading principle in many of the case studies that follow. Each of them is an attempt to locate the specific environmental objective in its appropriate physical and institutional 'environment', before the alternative regulation is proposed. Moreover, each proposal takes the goals of governments as its basic reference, thus relying on policy objectives as revealed in policy processes. In other words, each case study seeks consistency between problem, solution and instrument. Nevertheless the policy proposals in the following chapters are far from complete. In particular, legal aspects are missing. Of course, each proposal needs specific adjustments regarding the existing legal system in both the EU and the individual Member

States, but as economists, this is not what we had in mind in preparing this volume. Our effort precedes the concrete legal specification of environmental policy, that is, we focus on the existing incentive structure (current policies) and the way it might be changed in favor of the environment (our policy proposals).

Some Final Remarks

Concluding this introductory chapter, we summarize some of the most important points of departure and the jargon used in the following chapters. The focus is on instruments, assuming that the political process of goal setting is finished and has revealed a more or less accurate description of the desired future state of affairs. Hence, discrepancies between the policy goal and the social optimum are disregarded. In the case studies, the major evaluation criteria are environmental effectiveness and cost or economic efficiency. *Environmental effectiveness* is tested by comparing the (expected) results of instruments applied with objectives set by authorities, that is, environmental policy goals. If the policy goals could be attained at lower costs, *cost or economic efficiency* could be improved. If estimations of transaction costs related to the different forms of regulation are available, an attempt is also made to incorporate these costs into the notion of cost efficiency.

Other objectives of governmental policy, such as social stability or equitable income distribution, require that policy instruments also satisfy some criterion of *political acceptability,* which encompasses distributional impacts, transparency and concordance with existing institutional frameworks. The issue of distributional impacts is relevant, as it is a major determinant of acceptability of instruments by target groups who will have to cooperate in implementation. Each instrument affects the income distribution, which is a sensitive policy subject in many countries. Therefore, differences in distributional effects, for instance between a legal and an economic instrument or between two economic instruments, are a major consideration for the instrument choice. Transparency ensures that target groups have a clear view of the rationale for the introduction of the instrument and of the environmental purpose to be served.

Finally, the instruments proposed may interfere with the existing institutional context. Various institutional requirements could be distinguished. It is an advantage from a political acceptability point of view, if instruments are concordant with major institutional constructions, such as currently existing fiscal regimes. Further, EU regulations may imply restrictions with respect to the application of (both legal and economic) instruments by Member States, such as the prohibition of trade and competition distorting charges and regulations which require border control. Finally, some principles enjoy much political support in the EU, such as the 'polluter pays principle' and the 'subsidiarity principle'. Such principles receive special attention in the following chapters.

Changing incentive structures call for a well thought-out design of instruments. Each instrument requires specific adaptations regarding the environmental problem, while interferences with accepted principles or institutional frameworks have to be avoided. Otherwise, serious flaws are built into the regulation due to partial reasoning. For this reason problem orientation is the leading feature of Chapters 2 through 7. Each of these chapters explores one or more environmental problems and the possible use of economic instruments, given the current policy goals. In addition to the usual static approach to instrument choice, arguments of a more dynamic character are used frequently. At such moments of dynamic perspective, the initially available technological opportunities, the number of polluters, the scale of production and, subsequently, the scale of pollution, and even the initial policy goals may vary in the search for alternative incentive structures.

Besides, as already mentioned, environmental policy cannot and will not be an isolated activity of some governmental agency. Awareness is growing that effective environmental policies will hurt specific economic interests everywhere in the economy, although some will be hurt more than others. In this context Chapter 8 is devoted to the mutual interferences with non-environmental areas, especially fiscal policy. Finally, the last chapter of this volume summarizes and evaluates the policy proposals of Chapters 2 through 7 on their political practicability.

Notes

[1] Surveys of these methods can be found in Mäler (1985), Freeman (1985), Anderson and Bishop (1986), and Cropper and Oates (1992).

[2] Of course, from a social point of view the costs of pollution reduction are not a loss but an improvement, nor do the proceeds from the Pigovian tax or from selling pollution rights represent a loss to society as a whole.

[3] Note that this argument does not rely on any assumption about taxes or tradeable permits as being *optimal* Pigovian taxes or permits in that they precisely reflect values in the social optimum as discussed in the previous section (see also Baumol and Oates, 1988, pp. 160ff.)

References

Anderson, G.D. and R.C. Bishop (1986), 'The Valuation Problem'. In D.W. Bromley (ed.), *Natural Resource Economics; Policy Problems and Contemporary Analysis,* Kluwer Academic Publishers, Dordrecht, pp. 89-137.

Barthold, T.A. (1994), 'Issues in the Design of Environmental Excise Taxes', *Journal of Economic Perspectives,* 8, 1, pp. 133-151.

Baumol, W.J. and W.E. Oates (1971), 'The Use of Standards and Prices for the Protection of the Environment', *Swedish Journal of Economics,* 73, 1, pp. 42-54.

Baumol, W.J. and W.E. Oates (1988), *The Theory of Environmental Policy,* second edition, Cambridge UP, Cambridge.

Commission of the European Communities (CEC) (1992), *Towards Sustainability (Fifth Environmental Action Programme),* COM (92) 23, Brussels.

CEC (1992a), *The State of the Environment in the European Community* (accompanying document to the Fifth Environmental Action Programme), Brussels.

CEC (1993), *Growth, Competitiveness, Employment; the Challenges and Ways Forward into the 21st Century,* COM (93) 700, Brussels.

Coase, R. (1960), 'The Problem of Social Cost', *Journal of Law and Economics,* Vol. 3, pp. 1-44.

Crocker, T.D. and J.F. Shogren (1991), 'Preference Learning and Contingent Valuation Methods'. In: F.J. Dietz, F. van der Ploeg and J. van der Straaten (eds.), *Environmental Policy and the Economy,* North-Holland, Amsterdam, pp. 77-93.

Cropper, M.L. and W.E. Oates (1992), 'Environmental Economics: A Survey', *Journal of Economic Literature,* Vol. 30, Nr. 2, pp. 675-740.

Dales, J.H. (1968), *Pollution, Property and Prices,* University of Toronto Press, Toronto.

Dutch Council for the Environment (1992), *Vijfde Milieu-Actieprogramma van de Europese Gemeenschap* (Fifth Environmental Action Program of the EU), The Hague.

Faure, M., J. Vervaele, and A. Weale (eds.) (1994), *Environmental Standards in the European Union in an Interdisciplinary Framework,* MAKLU, Antwerpen.

Freeman, A.M. (1985), 'Methods for Assessing the Benefits of Environmental Programs'. In: A.V. Kneese and F.L. Sweeney (eds.), *Handbook of Natural Resource and Energy Economics,* Volume 1, North-Holland, pp. 223-270.

Hahn, R.W. (1989), 'Economic Prescriptions for Environmental Problems: How the Patient Followed the Doctor's Orders', *Journal of Economic Perspectives,* Vol. 3, pp. 95-114.

Hoehn, J.P. and A. Randall (1987), 'A Satisfactory Benefit Cost Indicator for Contingent Valuation', *Journal of Environmental Economics and Management,* Vol. 14, Nr. 3, pp. 226-247.

Liberatore, A. (1991), 'Problems of Transnational Policymaking. Environmental Policy in the European Community', *European Journal of Political Research,* pp. 281-305.

Mäler, K.G. (1985), 'Welfare Economics and the Environment'. In: A.V. Kneese and F.L. Sweeney (eds), *Handbook of Natural Resources and Energy Economics,* Vol. 1, North-Holland, Amsterdam, pp. 3-60.

Newberry, D.M.G. (1980), 'Externalities: The Theory of Environmental Policy'. In: G.A. Hughes and G.M. Heal (eds.), *Public Policy and the Tax System,* Allen and Unwin, London, pp. 106-149.

Pezzey, J. (1988), 'Market Mechanisms of Pollution Control: 'Polluter Pays', Economic and Practical Aspects'. In: R.K. Turner (ed.), *Sustainable Environmental Management,* Belhaven Press, London, pp. 190-242.

Pigou, A. (1920), *The Economics of Welfare,* Macmillan, London.

Smith, S. (1992), 'Taxation and the Environment: A Survey', *Fiscal Studies,* Vol. 13, Nr. 4, pp. 21-57.

Vollebergh, H.R.J. (1994), 'Environmental Taxes and Transaction Costs', *Tinbergen Institute Discussion Paper,* TI 94-96, Rotterdam.

Weale, A. (1994), 'Environmental Protection, the Four Freedoms and Competition among Rules'. In M. Faure, J. Vervaele, and A. Weale (eds.), *Environmental Standards in the European Union in an Interdisciplinary Framework,* MAKLU, Antwerpen, pp. 73-89.

Wildavsky, A. (1984), *The Politics of the Budgetary Process,* Little, Brown and Company, Boston.

2 Nutrient Emissions from Agriculture

An Alternative for the Current Abatement Policies in EU Member States

Frank J. Dietz
Heddeke Heijnes

Excessive Nutrient Use in European Agriculture[1]

Since World War II, agricultural production per hectare has increased considerably in Europe due to a rapid process of intensification of the production processes. Unfortunately, along with the increase in production of meat, milk, eggs, cereals, sugar beets, etc., environmental impacts from agricultural production processes have increased at the same pace, if not faster. To mention just some of the environmental problems, current agriculture substantially contributes to the dehydration of ecosystems, soil erosion and the contamination of soil, groundwater and surface water with pesticides and nutrients, such as phosphate and nitrogen.

High crop yields require high levels of inputs per hectare. Since fixed inputs (land quality, water supply) can only slowly be adapted for generating higher yields, high production levels usually coincide with high levels of variable inputs per hectare, such as fertilizers and pesticides (Tinker, 1988, pp. 15-16; CEC, 1989, pp. vii-x; Hoogervorst, 1990, pp. 218-219). Under these circumstances, however, higher levels of variable inputs reduce efficiency according to the law of diminishing returns. Since variable inputs are relatively cheap, their low efficiency has only marginal economic consequences for farmers. However, the nutrients which are not taken up by crops leach or evaporate into the environment, while the pesticides used cumulate as toxic substances in the environment depending on their persistence.

Large animal numbers per hectare result in large quantities of manure production per hectare. Most dairy farmers can, in general, use the slurry of their cattle on their own grassland. However, specialized pig and poultry farmers have only a limited amount of land available. For the discharge of manure they are generally dependent on the manure demand of arable farmers. But arable farmers usually prefer chemical fertilizers to provide the required nutrients, such as phosphate and nitrogen. Consequently, manure from the intensive livestock sector has become a waste product, being dumped on farmland rather than utilized. Especially in areas where intensive livestock farms are concentrated, manure dumping has serious environmental effects.

This paper deals with the environmental impacts of excessive nutrient use in EU agriculture. The issue is which mix of instruments would effectively and efficiently abate nutrient emissions from European agriculture. In the next section, first the environmental problems resulting from nutrient emissions will be described. In several EU Member States, policies have been developed to reduce the nutrient emissions from agriculture. The specific goals and the instruments used will briefly be discussed in the third section. Next, existing abatement policies are criticized for being ineffective and inefficient. Emphasis will be given to the inefficiency of the command-and-control instruments used. To improve the efficiency of environmental policy economists generally advocate the use of economic instruments, such as regulatory levies. In this respect several environmentalists and agricultural economists proposed a levy on the *use* of nutrients. The potentials of this form of price manipulation are discussed in the fifth section. In this case, however, an input levy has considerable allocative disadvantages, because of incentive distortions. As an alternative the introduction of a levy on nutrient *emissions* is advocated in the sixth section. The succeeding section is devoted to the European context of national initiatives for abating nutrient emissions. The EU will critically evaluate policy instruments used, especially from the perspective of the completion of the Internal Market. Furthermore, there is a case for a policy initiative on the EU level regarding the substantial international spillovers of nutrient emissions from agriculture. Finally, the last section offers some concluding remarks.

The Environmental Impacts of Nutrient Emissions from Agriculture

As far as figures are available, it is clear that in the EU the application levels of phosphate and nitrogen from manure and chemical fertilizers are higher than is necessary for the growing process of plants. This is indicated by the low efficiency rates of nutrient application, that is the percentage of the nutrients used that leaves the farm in agricultural products. For example, Schwarzmann and Von Meyer (1991, pp. 62-63) calculated a nitrogen efficiency of 23% and a phosphate efficiency of 35% for German agriculture. In Denmark the nitrogen efficiency rate is estimated at 30-35% (Dubgaard, 1991, p. 39) and in the Netherlands at approximately 20% (Olsthoorn, 1989; Centraal Bureau voor de Statistiek, 1989).

Low nutrient efficiency rates mean high losses of nutrients, implying nutrient leaching, that is, the emission of nutrients into the environment. Nitrate leaching, for instance, is a large-scale phenomenon in Europe, according to the calculations of RIVM and RIZA (1991, pp. 23-26), and RIVM (1992, pp. 74-77). The EU target value of 25 mg nitrogen per liter of perculation water is already exceeded in 65% of the agricultural area in Western Europe. For the limit value for drinking water (50 mg per liter) this percentage is 20%. Large-scale nutrient leaching has substantially contributed to at least three categories of environmental problems: eutrophication, groundwater pollution and acidification.[2] The nutrient flows related to agricultural processes are sketched in broad outline in Figure 2.1.

Nitrogen and phosphate emissions from agriculture contribute to the *eutrophication of surface water,* implying the excessive growth of algal and aquatic plants. In stagnant waters, such as lakes, slow-flowing rivers and canals, algal blooms can develop if concentrations of phosphorus exceed 0.15 mg per liter and if concentrations of nitrogen exceed 2.2 mg per liter. Hence, the critical load by weight for nitrogen is about 15 times higher than that for phosphorus. In most of these freshwater bodies the amounts of nitrogen available compared with those of phosphate do not exceed this ratio, so that phosphate is very often the nutrient that controls algal growth.

Phosphate application from both fertilizers and manure can, in principal, technically be attuned to crop needs. Excessive phosphate which is not taken up by plants, is largely absorbed by the soil, which acts as a buffer. Due to this buffering capacity, only a minor part of the excessively applied phosphate reaches the surface water. Currently, the major

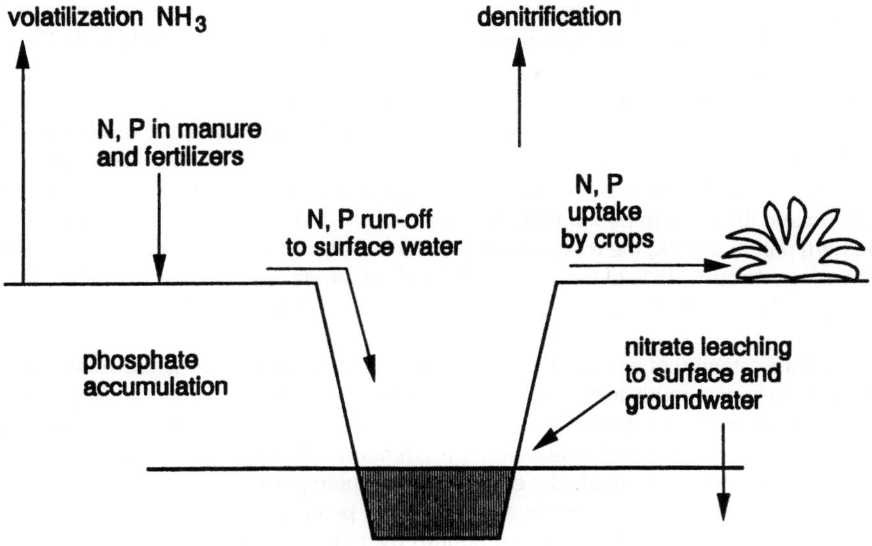

Figure 2.1: Phosphate and nitrate emissions from agriculture.

part of the phosphate pollution originates from raw sewage and from sewage works. However, the absolute and relative contribution of agriculture to the phosphate contamination will drastically increase in the near future for two reasons. Firstly, in countries such as France, the Netherlands, Germany and Denmark phosphate strippers are being installed in the sewage works at an increasing rate, which will considerably reduce the phosphate emissions from sewage. Secondly, the emissions from agriculture will increase appreciably since the continuing excessive use of phosphate rapidly fills up the buffering capacity of soils. As a result, more and more land will become satured with phosphate, producing a breakthrough of phosphate. This will ultimately result in the destruction of freshwater plant and fish life. Saturation of phosphate especially threatens areas with manure surpluses from intensive livestock farms. In the Netherlands, for instance, approximately 70% of the sandy soils were already found to be satured with phosphate (Willems and Hoogervorst, 1991, pp. 290-291; RIVM, 1993, p. 90).

In contrast to freshwater, algae growth due to eutrophication of estuarial, coastal and marine waters is mainly controled by the amount of nitrogen available. However, the control of nitrogen emissions from agricultural production processes is technically complicated. In addition to the uptake by crops, applied nitrogen disappears from the top soil through processes like denitrification and volatilization. The utilization of the nitrogen applied depends on local conditions, such as soil processes and rainfall. Since farmers aim for a maximum output, the uncertain efficiency of nitrogen applications have stimulated farmers to apply it in large amounts. The relatively low price of nitrogen probably encouraged farmers to consider inefficiently used nitrogen as a cheap insurance against disappointing yields (Shortle and Dunn, 1986; Hoogervorst, 1990). However, if the difference between the application and the uptake of nitrogen is not kept to a minimum, environmental degradation by eutrophication will result. The abundant algae growth in the Adriatic Sea along the Italian coast each summer since 1989, indicates the nightmare of large-scale eutrophication of marine waters.[3] In France, the Netherlands and Denmark concern for the eutrophication of marine waters is growing (Comolet and Pagnard, 1991, pp. 175-176; Van der Gaag *et al.*, 1991; Dubgaard, 1991).

A second category of environmental problems due to nutrient emissions is *nitrate pollution of groundwater,* jeopardizing the provision of drinking water. Approximately 70% of drinking water in the EU is produced from groundwater (CEC, 1989, p. 5). Due to high levels of nitrogen use in agriculture combined with the sensitivity of specific soils to nitrate losses, nitrate concentrations in groundwater continue to rise. Computations show that in 25% of the EU area nitrate levels may be expected to exceed the level of the EU drinking water standard of 50 mg/l (RIVM and RIZA, 1991, pp. 22-32). These areas are particularly found in the eastern and southern parts of the Netherlands, the western part of Denmark, the western part of Belgium, the northern and western parts of Germany, the southern part of the UK, the western part of France (Brittany) and the northern part of Italy (Po valley). Another 45% of the computed levels of nitrate leaching are between 25 and 50 mg/l, which is above the EU target value of 25 mg/l (RIVM and RIZA, 1991, pp. 22-32).

A third category of environmental problems is *acidification* due to the volatilization of ammonia from manure. In most EU Member States acidification from ammonia emissions is not perceived as a pressing environmental problem. However, in areas with large concentrations of (intensive) livestock farms ammonia emissions are substantial, such as in the western part of Denmark, the eastern and southern parts of the Netherlands and the western part of Belgium. Although acidification is perceived as a continental problem due to the dispersion of acidifying compounds over large distances, ammonia emissions have regional impacts.[4] For this reason ammonia in the Netherlands accounts for 65% of the deposition of acidifying substances that originate from Dutch sources, while ammonia emissions from manure contribute to 35% to the total acid deposition in the Netherlands (Bakema *et al.*, 1991, pp. 200-204; Erisman and Hey, 1991, p. 84).

In Europe the environmental impacts of overfertilization in general are intensified by a persistent trend towards specialization and concentration (particularly in France, Belgium, the Netherlands, the northern and western parts of Germany and Denmark). Specialization and concentration are induced by the search for economies of scale, which has resulted, for example, in an increasing animal production on fewer farms, in segregation of livestock and fodder production, and in short rotation of high yield crops which require intensive fertilization. Most intensive livestock farms have very little land available, which has two interrelated consequences. Firstly, the farms do not produce sufficient fodder for their livestock. In principle, cereals from arable farmers could be purchased, reducing at the same time the EU cereal surpluses. However, due to the EU price support for cereals, intensive livestock farmers prefer cheaper concentrated feedstuffs. These feedstuffs are mainly made up of tapioca, soya and maize gluten which are imported from outside the EU. Hence, huge amounts of nutrients are extracted from arable land elsewhere and shipped to the EU where they ultimately show up in the manure of millions of pigs and chickens.[5] Then, the lack of land has a second consequence: Intensive livestock farmers can impossibly use all these nutrients and, consequently, consider the manure as mere waste, to be disposed of in the cheapest way possible, implying manure dumping on their relatively small acreage of land.

The high concentration of intensive livestock farms in relatively small areas aggravates the environmental impacts of manure dumping. As an indication of this concentration Table 2.1 shows the numbers of pigs, chickens and cattle per hectare in seven EU regions. An additional indicator of the environmental pressure the intensive livestock sector brings to bear in these regions is the calculated phosphate production per hectare, that is the phosphate load per hectare originating from manure and calculated on the basis of the actually held number of animals.

Unfortunately, in some of the concentration areas most of the intensive livestock farms are predominantly located on sandy soils, as is the case in the Netherlands and

Belgium. In Denmark a transfer of (intensive) livestock farms is noted, coming from the heavy soils of the eastern Islands and going to the light (sandy) soils of western Jutland (CEC, 1989, Appendix p. 33; Dubgaard, 1989). Water movements through sandy soils are relatively rapid, which explains why leaching of nutrients to surface and groundwater is so high. In these areas high concentrations of intensive livestock farms coincide with a high sensitivity of the environment.

Table 2.1: Animal numbers and phosphate production per hectare in various EU regions in 1985.

Region	*Pigs*	*Chickens*	*Cattle*	*P_2O_5/ha*
Belgium	3.9	15.4	2.2	125
Brittany	2.8	45.9	1.6	108
Denmark	3.2	5.4	0.9	63
Lombardia	2.6	40.2	1.8	112
Netherlands	6.1	45.1	2.6	173
Northrine-Westfalia	3.8	7.5	1.3	84
Yorkshire/Humberside	1.5	9.4	2.0	96

Sources: Hoogervorst, 1990, p. 219; Mansholt, 1991, p. 345.

National Policies Concerning Nutrient Emissions from Agriculture[6]

The adverse environmental effects of nutrient leaching vary across the EU due to, among other things, differences in soil features. This partly explains the variety of current national abatement policies. In some cases no abatement policies exist because nutrient leaching is rather limited, such as in Greece, Portugal and Ireland. In large EU countries the environmental impacts of nutrient leaching are rather modest on the aggregated level of the whole nation, but quite strict in specific regions. In these cases specific abatement policies for agriculture are mainly initiated and implemented by regional authorities. In contrast, in small countries such as Denmark and the Netherlands nutrient leaching from agriculture is a national issue, resulting in national abatement policies. In the federal state of Belgium environmental protection is a regional competence, implying that Flanders and Wallonia are responsible for their own legislation and administration. Currently there are no regulations regarding the use of fertilizers and manure in Wallonia, although nutrient leaching is a substantial phenomenon in arable farming. Only the Flemish authorities have developed an abatement policy, mainly because most intensive livestock farms are located in Flanders.

Policies to abate nutrient leaching also differ because of differences in environmental pressure. Policy makers in Europe are primarily concerned with the contamination of groundwater with nitrates. Groundwater is by far the most important source of drinking water in the EU. In 1980, for reasons of public health, the EU established a directive containing the Maximum Admissible Concentration (MAC) of nitrates in drinking water which has been set at 50 mg nitrate per liter (Johnson and Corcelle, 1989, p. 48). Recent research indicates that considerable parts of the phreatic aquifers in Spain, France, England, Denmark, Germany and the Netherlands, which are of interest in terms of the water supply, contain more than 50 mg nitrate per liter (RIVM and RIZA, 1991, p. 29).[7] This implies that a substantial amount of the population uses drinking water containing too many nitrates, in particular when private wells are exploited (Baldock and Bennett, 1991, pp. 29-30). In most countries the concern for the rising nitrate contents of drinking water motivated policy makers to develop some form of policy for the abatement of nutrient leaching from agriculture. This concern also led to the promulgation by the European Commission of the 'Nitrate Directive' in 1991 (EC directive 91/676/EC, published in the Official Journal of the European Communities, 1991). This directive is meant to protect fresh waters, coastal waters and sea water from nitrate originating from non-point agricultural sources. To this end, Member States must identify vulnerable zones in which agricultural pollution endangers the aquatic environment and, subsequently, have the obligation to prevent applications of nitrogen that exceed 170 kg per hectare in these zones by the end of the year 1999.[8]

Eutrophication from agricultural nutrients has a lower priority in current abatement policies than groundwater pollution. In general, most attention is focused on eutrophication of coastal waters which is prejudicial to both tourism and marine activities. Moreover, the North Sea countries have agreed to considerably decrease the emissions of nutrients. As a result some measures have been taken by national and regional authorities varying from a pure curative nature (for example, mechanical elimination of algae along the coastline in Brittany) to a more or less preventive character (attempts to reduce the nutrient emissions in, for example, Denmark and the Netherlands; see for a description of specific measures the next paragraphs). Concern for the eutrophication of fresh waters with phosphates from agriculture is, generally speaking, confined to a few countries (the Flemish region in Belgium, Denmark, the Netherlands, United Kingdom). A central issue in these countries is that at the moment only a minor part of the fresh water eutrophication can be ascribed to agricultural activities. The major part stems from the discharge of sewage. As indicated in the previous section this will change rapidly in the near future due to the current farming practices.

In most countries the contribution of livestock farming to the acidification is too low to alarm policy makers. Only in Denmark and the Netherlands are agricultural ammonia emissions considerable (up to one third of the acidifying depositions in the Netherlands), which has led to specific policy measures.

The current policies for the abatement of nutrient emissions from agriculture differ widely among the EU Member States. In the United Kingdom, for example, farmers are asked and advised to volunteer for changes in their farming practices. They are expected to act according to the code of 'good agricultural practice'. Manure storage capacity for only four months is mandatory and discharges of slurry are prohibited. In Italy a policy concerning the application of fertilizers is currently being developed. In addition, in Lombardia and Emilia Romagna the spreading of manure is restricted by application standards expressed in kilograms of meat and specified for different categories of animals.

The regulations in France and Germany are more stringent, that is the regulations in the 'Departements' of Brittany and those in the 'Länder' Northrine-Westfalia, Schleswig-Holstein and Lower Saxony substantially restrict farming practices. Manuring and fertilizing standards for nitrogen have been imposed, storage capacity for manure is obligatory (from 45 days in the whole of France up to 6 months in Brittany) and manure spreading is not allowed in winter (although the actual non-spreading periods vary considerably).

The most stringent policies are to be found in Flanders, Denmark and the Netherlands. However, substantial differences must be noted. Manuring standards have been commonly imposed, but at considerably different levels and measured in various units (kg phosphate per ha in the Netherlands, kg nitrogen and phosphate in Flanders and the manure equivalent of 2.3 animal units in Denmark). The mandatory storage capacity for manure varies from two to six months in Flanders to nine months in Denmark. Further, regulations to decrease ammonia emissions exist, such as the sealing of stables and manure storages and the obligation to plough the spread manure into the soil within 12 hours (Denmark), 24 hours (Flanders) or 36 hours (Netherlands). In both the Netherlands and Flanders manure banks are established which mediate in the sale and transport of manure from surplus areas to 'shortage' areas. Dutch farmers, financially supported by the government, are undertaking the ambitious project of developing a network of large-scale reprocessing plants for surplus livestock waste. This initiative may also be pursued in Flanders and Brittany.

In all EU countries regulations concerning the application of chemical fertilizer are much less strict than those on manure application. In most cases farmers are only advised to decrease the use of, in particular, nitrogen fertilizer. In Denmark farmers with at least ten hectares of agricultural land are obliged to prepare a fertilizer management plan. In most countries farmers face additional restrictions on nutrient use in drinking water collection areas and nature protection zones, for which they are financially compensated in most cases.

Generally speaking, the current policies for abating nutrient emissions from agriculture in EU Member States almost exclusively consist of direct regulations, usually based on some form of legislation combined with sanctions for non-compliance. Financial incentives, or economic instruments as they are often called in the literature, such as levies and transferable pollution rights, are rarely used. In the sporadic cases that a levy is imposed it is meant, in the first place, to raise revenues to finance policy initiatives, such as the development of industrial manure processing plants in the Netherlands,[9] the establishment of manure banks in Flanders and the improvement of surface water in Spain. Subsidies are more often used to influence the behavior of farmers by way of, for example, subsidies for the construction of storage capacity.

The Failures of Current Policies

In this section current policies in different EU Member States to abate nutrient leakages are shown to fail with respect to both its effectiveness and efficiency. We consider abatement policies *effective* if environment-damaging nutrient emissions from agriculture are eliminated. Hence, ecological sustainability of agricultural processes is the evaluation criterium for the specific goals of current abatement policies. In addition, we consider current abatement policies *efficient* if the environment-damaging nutrient emissions are eliminated at the lowest social costs.

To operationalize the objective of sustainability for nutrient use in agriculture, standards must be developed for either nutrient applications or nutrient losses. The goals of current abatement policies - if there are any - usually take the form of a set of application standards. There are several problems with the existing standards. To begin with, the abatement policies only partly cover the use of nutrients in agriculture. In most cases attention is only given to nitrogen leakages, disregarding the eutrophication caused by phosphates. Only the Dutch government regulates the use of phosphate, allowing at the same time, however, the free use of nitrogen.[10] Moreover, the standards for the maximum application of nutrients per hectare are, on the whole, far too high to be consistent with the carrying capacity of nature, that is the capacity of nature to process nutrients lost from agricultural activities in a sustainable way.

Apart from drinking water catchment areas, legislative control of maximum chemical fertilizer use does not exist in EU countries. In contrast, regulatory requirements for farmers regarding animal manures are far more extensive throughout the Community. What is peculiar, however, is the exclusion of various categories of animals in policies, as is the case in the Netherlands. Until 1992 the regulations were only confined to cattle, pigs and poultry. As could be expected, more and more farmers bypassed such regulations on manure by changing over to ducks, geese and, especially, sheep. Since January 1992 these categories (and also fur-producing animals) have been included, but there are still categories that are free from regulation, such as horses and roes.

In the next decade the differences in goals between the nutrient policies of EU Member States could decrease, while at the same time increasing their effectiveness, due to the implementation of the EU directive 91/676/EC (the Nitrate Directive). This directive compels Member States to control, stepwise, the water pollution originating from agricultural sources. In 1993 all countries had to present a list of vulnerable areas in their country. Each Member State has a period of two years to formulate action programs containing measures to reduce the use of nitrogen from both manure and chemical fertilizers. These action programs must be revised every four years. The first action program has to be carried out in the four years following the formulation of the program (1995-1999). In this period a maximum application of 210 kg nitrogen per hectare is allowed. For the next four years (1999-2004) a strict standard of 170 kg nitrogen per hectare has already been announced.

In short, the currently used sets of application standards - being the goal of existing legislations - are incomplete. In addition, the allowed application levels considerably exceed the carrying capacity of nature. In principle, these failures could quite easily be corrected. Basically, what is needed is the introduction of environmental standards for the maximum application of nutrients, covering *all* nutrients leaching from agriculture.[11]

The next step is much more difficult: to design an efficient policy that will encourage farmers to comply with the standards that have been set. The core of the policies currently in effect is a set of command-and-control instruments, including various prohibitions and prescriptions at farm level, such as an animal/hectare ratio, a ban on manure spreading beyond the growing season, obligations to plough spread manure into the land in order to prevent volatilization of ammonia, obligations to build accommodation to store manure beyond the growing season, special restrictions concerning farming practices in drinking water catchment areas, etc. This command and control approach seriously limits the options within individual farming practices, especially the option of how to reduce nutrient losses.

Nutrient leakages vary greatly depending on specific local circumstances, which limit either the effectiveness or the efficiency of generally prescribed abatement actions. A set of uniform prohibitions and prescriptions which adequately prevent the washing-out of nutrients on one type of soil (e.g. clay) may prove inadequate on another type of soil (e.g. sand). If the set of prohibitions and prescriptions is made so restrictive that washing-out is prevented on all types of soil, the set will appear to be inefficient, because less restrictive measures, and hence lower costs, will suffice at locations with lower rates of washing-out.

To prevent such inefficiencies in the abatement of nutrient leakages, specific local conditions (such as the type of soil, the level of the groundwater, the type of crop that has to be grown, the amount of rainfall, etc.) must be taken into account. Consequently, a tailor-made set of measures is needed for each type of location which must be fit to the specific circumstances of nutrient leaching. The accumulation of detailed and specific measures will result in a very complicated system of regulations, as is already the case in the Netherlands (Dietz and Hoogervorst, 1991; Dietz and Termeer, 1991), in Flanders (De Clerq and Sennesael, 1991) and in Denmark (Dubgaard, 1991). Apart from the costs incurred in gathering information about local circumstances as well as those incurred in enforcing all prohibitions and prescriptions, the command and control structure will also have paralyzing effects on the agricultural sector. The prohibitions and prescriptions greatly restrict farming practices, causing resistance from individual farmers and their organizations.

Apart from the high costs of gathering information and maintaining the regulations (= agency costs) as well as apart from the paralyzing effects of the complicated command and control systems used, there is no incentive to decrease nutrient losses below the actual level of the standards imposed. As mentioned before, if any standard for applying maximum levels of nutrients is present, it does not correspond with a sustainable use of nutrients. In some cases (e.g. Flanders, the Netherlands) the government intends to decrease the allowed manuring and fertilizing levels step by step, until ecologically sustainable standards are ultimately reached. But it is not yet clear what these 'end-state standards' will look like and when they will have to be met. In the meantime, farmers who comply with the prevailing standards (which are far too high from an environmental point of view), have no incentive for decreasing their nutrient losses any further in the direction of sustainable levels.

To summarize, the existing abatement policies concerning excessive nutrient use in European agriculture are ineffective, because standards for nutrient use - if present - considerably exceed the carrying capacity of nature. Moreover, the predominant choice for command and control instruments for the abatement of nutrient leaching seems a rather inefficient approach. In order to improve cost efficiency, environmental economists urge the use of economic instruments in policy.

An Input Levy on Nutrients?

Arguments in favor of the use of economic instruments in environmental policy originate from welfare economics. If applied to the case of nutrient emissions from agriculture, nutrient losses would be seen as an externality problem. Farmers using excessive amounts of nutrients do not take into account the increasing production costs of other producers and the decreasing utility levels of consumers due to the environmental degradation caused by nutrient emissions. As Pigou (1920, p. 192) has already recommended, the activity causing the negative externality must be taxed at such a level that net social benefits will be maximized. However, optimization meets with fundamental problems, since it appears to be impossible to calculate *ex ante* the tax level which equals the marginal social costs and the marginal social benefits of decreasing nutrient losses. In particular estimation of the (marginal) benefits meets with insurmountable problems due to the dynamic nature of the problem of nutrient leaching.[12] So long as the marginal benefits of decreasing nutrient losses for current and future generations cannot be remotely estimated, "optimality is not a policy option" (Hanley, 1990, p. 136).

This conclusion does not imply that economic instruments are useless for controlling nutrient leaching. As Baumol and Oates (1971; 1988, pp. 159-176) demonstrated, a set of economic instruments can, in theory, achieve exogenous policy targets, such as emission levels, at lower total resource costs than uniform regulation. Economic instruments permit flexibility in the amount of pollution reduction achieved by each source, allowing polluters with low abatement costs per emitted unit to reduce emissions to a greater degree than polluters with high abatement costs per emitted unit.

One such option is a levy on the *use* of nutrients. This levy is most popular among agricultural economists and environmentalists, judged by the attention given to the input levy in the literature. Levies on the input of chemical fertilizers have been studied frequently (cf. England, 1986; Dubgaard, 1989; Hanley, 1990; Von Meyer, undated). Most of the studies only deal with nitrogen use, but could easily be extended to the application of phosphate and, if desired, to other nutrients. In addition, as far as we know, taxation of concentrated feedstuff has not been considered despite its high nutrients content. There is, however, no reason for exempting this category of nutrient inputs. Apparently, studies of this kind are exclusively inspired by the problem of excessive nutrient use on arable land and grassland.

The basic idea of nutrient price manipulation is to discourage the use of nutrients by a considerable price increase relative to that of the output, i.e. crops including grass.[13] This is likely to have two effects: firstly, a reduction in the optimum rate of nutrients for each individual product, and secondly, a substitution effect at farm level away from products demanding high levels of nutrients (England, 1986, p. 14).

Although improvements in nutrient efficiency are anticipated in the literature, an input levy on nutrients as an instrument to abate nutrient leaching has some major disadvantages. A uniform levy on nutrients in fertilizers and concentrated feedstuffs does not differentiate between nutrients being utilized in farming products and nutrients being lost into the environment. This is no major problem if the loss of nutrients varies proportionally with the degree of intensification. Then, high levels of nutrient use correspond with high environmental pressure and vice versa. But as De Wit (1988) points out, this proportionality is not a matter of course in practice. This implies that a farmer who farms intensively but uses nutrients efficiently is penalized more than a farmer who farms extensively and uses nutrients less efficiently. Moreover, Dubgaard (1989) points out a similar unintended and undesired effect of - in his particular study - a levy on nitrogen fertilizer: the least polluting but highly nitrogen-demanding crops (winter wheat and oilseed rape) are penalized more heavily than crops demanding less nitrogen but from which leaching is substantially greater (such as peas and spring cereals). These studies indicate that marginal damage costs are not equal across arable activities. The same holds for the intensive livestock sector. A uniform levy on concentrated feedstuffs does not impose a higher penalty on farmers located on the vulnerable sandy soils than on farmers located on the less vulnerable clay soils. A differentiation of the level of the levy is required in order to avoid these undesired effects.

Levy differentiation in this case could take the form of the allocation of exemptions from the levy, corresponding with the carrying capacity of nature per hectare in a specific area. However, this option meets with at least two major problems. Firstly, such an exemption will be conceived as a levy-free right on nutrient use. Farmers may be tempted to trade levy-free nutrient rights, a result which is feared by the European Commission (CEC, 1989, p. 120). In the areas with intensive forms of agriculture the input levy implies a substantial cost increase, which farmers will attempt to avoid by buying levy-free nutrient rights from collegues in less vulnerable areas which, in addition, could be dominated by extensive forms of agriculture. A concentration of levy-free nutrient rights is likely. As long as the price of these rights (including the transaction costs) is lower than the marginal tax rate on nutrient use, thwarting of the environmental policy by ongoing concentration can be expected. Secondly, if such concentrations are considered to be undesirable, a prohibition on the trading of levy-free nutrient rights is required. However, such a prohibition would imply a huge administrative task for the regulating agency, including the monitoring of the actual nutrient use per hectare.

The last objection to some form of input levy on nutrients dealt with here is the absence of any incentive to avoid nutrient leaching once the nutrients are bought. In intensive farming a levy on nutrients in purchased animal feed would considerably increase the nutrient efficiency in fattening pigs and chickens. Consequently, manure will contain fewer nutrients than presently. But farmers have no direct incentive to deal carefully with the remaining nutrients in manure. The prevention of nutrient leaching by, for example, the careful storing of manure beyond the growing season, means high additional costs for the farmer. On the other hand, farmers will probably receive a better price for nutrients from manure due to the price rise in fertilizers. However, arable farmers will not be willing to pay the same price for nutrients from manure as for nutrients from fertilizers, because crops take up fewer nutrients from manure than from fertilizers (cf. CEC, 1989, p. 37). Also, arable farmers are reluctant to buy manure as long as the nutrient content cannot be guaranteed, and manure contains heavy metals and seeds from weeds.

To summarize, a uniform input levy on nutrients has substantial allocative disadvantages, while a differentiation in the levy rate invokes substantial agency costs. The most serious objection is, however, that the damaging emissions themselves are not taxed. Consequently, incentives are lacking to avoid or to reduce nutrient losses once the nutrients are bought. To put it differently, the incentive structure does not fit the problem at hand.

An Emission Levy on Nutrients as an Alternative

At first sight it seems that the incentive structure could be improved by simply imposing a levy on the emissions causing environmental deterioration. However, this requires that the taxing agency is familiar with the character and amount of the emissions as well as the locations of all nutrient emission sources, which encompasses in this case the total agricultural acreage in the EU. Hence, the agency costs of a levy on nutrient emissions would be high. This explains to a large extent the preference for the input levy in the literature regarding its relatively low agency costs. But by doing so the allocative advantage of the emission levy is sacrificed too early. It appears to be possible to reduce the agency costs considerably without giving up the principle of the emission tax.

The agency costs could be substantially reduced if the taxing agency does not try to measure the nutrient *emissions,* but farmers keep account of their nutrient *losses.* The principle is simple and is similar to bookkeeping. Each farmer keeps track of the nutrients which enter his firm in the form of, for example, chemical fertilizers, concentrated feedstuff, fodder or manure. He also counts the nutrients leaving the farm in products, such as meat, milk, eggs, crops and manure. The credit balance between incoming and outgoing nutrient streams is the *nutrient loss* of the farm.

Nutrient accountancy helps a farmer track down nutrient dissipation. This information could stimulate him to use nutrients more efficiently, at the same time reducing the environmental pressure. However, awareness of nutrient dissipation is unlikely to change the farmer's behavior, due to the relatively low price of nutrients. On the contrary, as noticed earlier, the relatively low price of, for example, nitrogen tempts farmers to use it excessively as cheap insurance against disappointing yields. Changing this behavior requires at least a substantial increase in the costs of nutrient losses.

Concerning the cost increase of nutrient losses by imposing a levy, a complicating factor is, however, that in most cases individual losses of nutrients as determined by nutrient accountancy are not equivalent to nutrient emissions. Phosphate losses do not immediately lead to emissions due to the buffering capacity of the soil. Nevertheless, filling up this buffering capacity must, in our view, be interpreted as environmental pressure, because, after laying fallow, it takes many years before the buffering capacity of phosphate satured soils have been restored. Keeping the phosphate combining capacity of the soil stable is, therefore, a necessary condition for sustainable agriculture, requiring the prevention of phosphate losses from agriculture. Phosphate losses can almost entirely be prevented, because the phosphate dose can quite accurately be attuned to the crop needs. In this context we define the *phosphate surplus* as the positive difference between the used amount of phosphate and the uptake by the crop.

It is much more difficult to adapt nitrogen doses to crop needs. Crops only take up part of the nitrogen present in fertilizers or manure. This uptake varies across crops and seasons. Whether the remaining nitrogen damages the environment depends on specific local circumstances, such as type of soil, rainfall, level of groundwater, denitrification, winter coverage, deposition of nitrogen compounds originating from other sources, etc. Thus, not all nitrogen losses lower environmental quality, for the capacity of nature to process nitrogen varies across locations. Nitrogen losses are damaging to the environment to the extent that the neutralizing capacity of nature is exceeded. In this context we define the *nitrogen surplus* as the positive difference between the nitrogen losses of a farm and the amount of nitrogen nature can sustainably neutralize at a specific location.[14] The phosphate and the nitrogen surplus together are the *nutrient surplus.*

To summarize, the nutrient loss is the difference between the input and output of nutrients at farm level. The nutrient surplus is, subsequently, the difference between nutrient losses and harmless emissions. Which parts of nutrient losses are damaging and which parts are not, is especially uncertain for nitrogen. In the ideal case the damaging nature of nitrogen emissions is measured on each lot. Although technically feasible, measurement of the nutrient surplus on each lot appears to be costly. Besides, facilities for taking and analyzing all the soil samples needed are lacking as yet. Thus, for the time being, the nutrient surplus has to be calculated on the basis of average circumstances for, and impacts of, nutrient leaching. These average circumstances and impacts can be translated into a set of fertilizing and manuring standards, termed *environmental standards* to distinguish them from the standards used in farmers' advisory services and which are based on maximum output levels without any environmental considerations. Environmental standards specify the maximum amount of nutrients that can be applied per hectare per year without causing damage to the environment, taking into account the type of crops grown and the type of soil present.

Several attempts have been made to develop such environmental standards (e.g. Dietz and Hoogervorst, 1989; 1991; Goossensen and Meeuwissen, 1990). For the Netherlands the state-of-the-art knowledge is summarized in Table 2.2. These environmental standards are derived from the maximum losses of nutrients to soil, water and air that are acceptable from an environmental point of view. These maximum nutrient levels encompass all manures and fertilizers. The environmental standards for application of phosphate are equivalent to the phosphate uptake by agricultural crops on different types of soil. The environmental standards for the application of nitrogen are based on acceptable emissions of ammonia, on the European target level for nitrate in drinking water, as well as average circumstances regarding the nitrogen deposition, denitrification, etc. A farm has a nutrient surplus if the *calculated* amount of nutrients used on the land (using the accountancy system) is larger than the maximum nutrient load per hectare that may be applied according to the environmental standards of Table 2.2.

Table 2.2: Environmental standards for the use of phosphate and nitrogen.

	Grassland	*Arable land*	*Green maize*
Phosphate (kg P_2O_5/ha)	110	70	75
Nitrogen (kg N/ha)			
- clay soil	300	100	100
- sandy soil	250	100	75

Source: Ministry of Agriculture, 1990; RIVM, 1993.

The standards in Table 2.2 are the goals of what we call a nutrient policy. The remaining issue is how to achieve these goals in the most efficient way. A bald announcement that the environmental standards will be effective as from a certain date in the future and that they will be enforced if not complied with, will not work. Regarding the high level of nutrient leaching, farmers must be stimulated to modify their farming practices to an ecologically sustainable level long before the environmental standards have to be attained. In this context, Dietz and Hoogervorst (1989; 1991) propose using the nutrient surplus as a tax base for a regulatory emission levy. Imposing a levy on the surpluses of phosphate and nitrogen acts as an incentive to farmers to prevent nutrient losses. The surplus levy confronts farmers with an additional trade-off. On the one hand money can be saved by reducing nutrient use, which decreases the losses of nutrients and, subsequently, avoids the costs of the surplus levy. On the other hand, farmers may fear a considerable reduction of output from a substantial reduction in the nutrients used, which would jeopardize income generation and outweigh the avoided costs of the surplus levy. However, several studies (Tinker, 1988; Meyer and Lalkens, 1988; Neeteson, 1989) suggest that considerable reductions in nitrogen application can be achieved with little or no reductions in crop yields. This result is not surprising regarding the low nutrient efficiency figures mentioned in Section 2.

The most important gain of the surplus levy arrangement is that, in contrast to the currently dominant command and control instruments, the individual and diffuse knowledge about the specific physical conditions of farm and lot is mobilized in favor of the environment. For decades this knowledge has been used for producing as much (cereals, milk, meat) as possible, ignoring the loss of nutrient inputs in the hunt for higher production levels. The introduction of nutrient accountancy will make farmers aware of the, on average, substantial losses of nutrients. In addition, the surplus levy raises the costs of nutrient leakages considerably, which stimulates farmers to look for ways of reducing their nutrient surplus. There are numerous ways to decrease the nutrient surplus: reduction in the use of chemical fertilizers, substitution of manure by fertilizers, use of soil coverage during the winter season, reduction of the nutrient contents of concentrated feedstuff, transportation of manure to farmers who need nutrients, delivery of surplus manure to an industrial manure processing plant, etc. The choice of one or more methods of nutrient surplus reduction depends on the local physical circumstances, with which the farmer is most familiar (or which the surplus levy will encourage him to explore).

Dietz and Hoogervorst (1991) calculated for the case of the Netherlands that a levy of 1.25 guilders per kg phosphate surplus and 1.25 guilders per kg nitrogen surplus is needed to give the surplus levy an allocative function. It is difficult to compare the level of this surplus levy and the level of the existing ear-marked levy on manure production, because the basis of the levies differs. The latter has no regulatory target and is only meant as a source to finance abatement initiatives, such as the development of manure processing plants. As a rough indication it can be assumed that the level of the surplus levy is approximately 10 times higher than that of the existing levy. The surplus levy will also increase the price of chemical fertilizers leaching into the environment. Nitrogen or phosphate emissions from fertilizers will cost a farmer 1.25 guilders per kg, which will increase the price of leached fertilizers by approximately 200%.

A major objection to the introduction of the surplus levy could be the costs of collecting the necessary information. In particular the need to determine the nutrient surplus for each of the hundreds of thousands of farms would be very expensive for the taxing agency. On further consideration these costs appear highly dependent on the way the surplus levy arrangement is organized. The usual approach in fact induces high costs. To start with, an exact delineation of the clay and sandy soils is required. A lot of farms are located on mixed soil types, which implies that using standards developed for pure clay or pure sand will either harm the environment or mean an unfair increase in production costs for the farmer. Furthermore, farmers may claim that specific local circumstances (low rainfall, low groundwater level, low deposition) and special measures taken on the farm (winter coverage, manure injection) have made the *actual* nutrient surplus much lower than the *calculated* nutrient surplus. If all these basically reasonable claims must be checked (by the agency who taxes the nutrient surplus or by the authority for appeal), agency costs will rise prohibitively. Also, the character of the policy would change and would come to resemble the current situation: farmers feeling dependent on regulations they see as incomplete and unfair, excessively limiting their choice of farming practice. If this were the case, abatement policy would suffer severely from the non-compliance of farmers.

In our view the information costs for the taxing agency can be decreased considerably, while at the same time the compliance of farmers can be improved, if the burden of proof is put on the individual farmer. Then a farmer who claims a lower nutrient surplus than is calculated, has to demonstrate this, for instance, on the basis of an analysis of soil samples by an independent laboratory. Of course, shifting the burden of proof will also shift the costs of proof to the farmer. However, if these costs are not excessive (or could be reduced),[15] it may be expected that this institutional arrangement really will motivate farmers to comply with a system of nutrient accountancy and the determination of the individual nutrient surplus, because most decisions concerning farming practice are left to the farmer.

Another major objection that could be raised against the surplus levy arrangement concerns enforcement. It appears initially that a levy based on nutrient accountancy of individual farmers would be extremely susceptible to fraud. Farmers are continuously tempted to 'reduce' their nutrient surplus by registering less fertilizer use and less manure spreading than is actually the case and by recording a lower nutrient content of manure than is factual. Fortunately, the parties involved in nutrient transactions have incompatible interests. Arable farmers, (intensive) livestock farmers, dairy farmers, producers of fertilizers and concentrated feedstuff, manure banks, manure processing plants, all have to register their inputs and outputs of nutrients. The interest of, for example, an intensive livestock farmer supplying manure to an arable farmer is to register as high a nutrient content as possible for the manure delivered, while the interest of the arable farmer is the registration of a nutrient content as low as possible. If none of the farmers has power over the other, the manure transaction will only be effected if they both agree on the nutrient content, which will be very close to, if not exactly the same as the actual nutrient content. In general, it is essential for imposing a levy on the nutrient surplus that the (nutrient) gain of the one contracting party is the (nutrient) loss of the other party. This is not to say that malversations can be excluded.[16] But the incompatible interests of contracting parties implicitly introduce mutual checks into the system, reducing considerably the agency costs.

To summarize, an effective nutrient policy requires a set of sustainable application standards for nutrients from both manure and chemical fertilizers. The transition of current practices of manure dumping and overfertilization into a more or less sustainable use of nutrients seems most efficiently effected by imposing a levy on the nutrient surplus of individual farms, supplying industries and processing companies. In the literature a levy on the input of nutrients is often advocated. However, an input levy will generate substantial incentive distortions. On the other hand, imposing a levy on the nutrient surplus could generate high agency costs. But the agency costs can be kept low if the burden of proof is placed on the nutrient holder. As a result the transaction costs of nutrient holders (information about nutrient contents of manure and products is needed, negotiations concerning nutrient transactions) will increase.

The European Context of National Abatement Policies

So far the EU context of the current national abatement policies on nutrient leaching has hardly been mentioned. In light of the current political priority, the completion of the Internal Market, the European Commission is especially interested in critically evaluating existing regulations as well as additional measures or alternative instruments, such as economic instruments. The complicated legal aspects will not be dealt with in this section. Here, our modest intention is to shed some light on the EU pitfalls for both the strategy of direct regulation which the EU Member States with some sort of nutrient policy have opted for so far, and the alternative strategy of employing considerably more economic instruments, such as an input levy on nutrients or a levy on a nutrient surplus.

According to the principle of subsidiarity of Article 130R of the EU Treaty, environmental policy has to be formulated and carried out at the lowest possible administrative level. Thus, without international environmental spillovers, national authorities have to deal with the problems of nutrient leaching. The subsidiarity principle implies that one country may have a stricter abatement policy than other EU Member States, because of regionally differentiated environmental conditions. The Dutch government, for example, could do so because the nutrient leaching problem in the Netherlands is much more severe than in other EU countries.

However, this degree of freedom is restricted by Article 100A of the Treaty, which prohibits the national environmental authority from imposing arbitrary or hidden trade barriers. The issue here is to what extent environmental policy generates market segmentation arising from border controls and from market entry barriers (Siebert, 1991). For instance, on this basis the European Commission raised doubts as to whether the Dutch government subsidies up to 35 per cent of the development and construction costs of manure processing plants, are compatible with the Treaty of Rome (Official Journal of the EU, September 1990). Finally, the Commission announced in December 1990 that this subsidy was permissible. The fact that about half of this subsidy is financed by the farmers themselves (by means of the revenue raised from the excess levy) was instrumental in the decision. The Commission nevertheless emphasized that there could be no question of subsidizing the operation of the manure processing plants, a subsidy which Dutch farmers had tacitly hoped and also lobbied for. The Commission announced that further proposals in the Netherlands to control nutrient leaching from agriculture will be closely examined (Bennett, 1991).

The case of the Dutch subsidy for industrial manure processing illustrates the growing concern in Europe about the adverse effects of (national) environmental regulations on fair competition. Generally speaking, the core of most regulatory approaches in EU Member States is a licensing procedure, creating market entry obstacles which protect the existing firms against the competition of newcomers, as long as licences are not transferable (which is normal practice). To avoid segmentation of the European market, national licensing procedures must be harmonized. However, harmonization requires a European-wide definition of emission standards - either uniform or differentiated for specific regions - which are expected to be met by the application of the currently known abatement technologies. Instead of the European authorities defining the (continuously changing) currently known abatement technologies, it is clearly much more efficient to encourage individual firms to search for new technological solutions which meet environmental standards. As Siebert (1991) demonstrates, economic instruments reduce the role of regulatory procedures and thus make market entry easier. From the Single Market perspective, economic instruments therefore have the appeal of reducing market barriers and segmentation.

Unfortunately, some economic instruments can also generate market segmentation in Europe. In this respect, proposals for an input levy on nutrients will meet with far more objections than the emission levy on nutrients, such as the levy on a nutrient surplus. In particular the requirement that different environmental conditions have to be taken into account by the allocation of exemptions from the levy (implying the allocation of levy-free rights on nutrient use), means additional controls at the borders between Member States. This could be interpreted as conflicting with Article 100A. If not, it is in any case incompatible with the aim to decrease tariff and non-tariff barriers in light of the completion of the Internal Market. Moreover, the introduction of a differentiated input levy on nutrients could also hamper fiscal harmonization in the EU, because such a levy is seen as an indirect tax.

A levy on the emission of nutrients into the environment does not have these disadvantages. The base of the levy, the nutrient surplus, is the same for all Member States. If trading in nutrients (e.g. manure from surplus regions to shortage regions) were to develop, border controls would not be required as is the case with the input levy. As far as we know, the European fiscal harmonization program does not apply to emission levies. But even if it did, a uniform levy would suffice, because the different environmental conditions do not need to be expressed in tariff differentiation.

So far nutrient leaching has been discussed as a purely national problem which needs to be abated at a national level in line with the subsidiarity principle. Indeed, part of the nutrient problem concerns purely national, or often only regionally felt environmental effects, such as nitrate pollution of groundwater and the eutrophication of streams and lakes. However, a substantial partition of the leached nutrients has impacts which are internationally felt, such as the pollution of groundwater in border areas, the eutrophication of rivers and marine waters, like the North Sea and the Mediterranean Sea, and acidification due to ammonia emissions. These international spillovers demand an international approach. The North Sea conference of 1990 was a step in this direction. However, immediate results cannot be expected because national policies are as yet still dominated by the fear of the national agricultural sector losing international competitiveness if a strict national abatement policy were to be implemented.

The brief discussion above warrants the conclusion that the introduction of a regulatory levy on the nutrient surplus of individual farms as advocated in the previous section, need not contravene existing European regulations concerning fair competition nor interfere with the aim of a single European market. An input levy on nutrients and, in particular, command-and-control instruments could generate substantial barriers to national markets. Such barriers will meet with strong opposition from the European Commission, which fears the slowing down of the European integration process.

Concluding Remarks

In agriculture, excessive nutrient use is normal practice in most EU Member States. The resulting nutrient emissions cause serious environmental problems, especially in areas where intensive livestock farms are concentrated. The environmental effects of nutrient leaching generally have a regional or a national character, which makes European initiatives for abatement at first sight unnecessary. As far as regional or national abatement policies exist, they fall short in both effectiveness and efficiency. The effectiveness is poor because the use of nutrients is only partly covered (the extensive application of phosphates in particular is largely neglected) and because much higher nutrient application levels are allowed than nature can carry sustainably. To improve the effectiveness of abatement policies environmental standards need to be introduced, indicating the maximum application of nutrients per hectare (at least differentiated per type of soil and type of crop grown), covering all nutrients leaching from agriculture.

The efficiency of existing abatement policies is poor because regulating authorities have unilaterally chosen for command-and-control instruments. If the regulating authorities would like to take into account the specific local conditions of nutrient leaching, tremendous costs will be incurred for collecting information and for enforcing all prohibitions and prescriptions. Such a tight command-and-control approach will also have paralyzing effects on the agricultural sector.

The efficiency of controling nutrient leaching could be considerably improved by introducing economic instruments. In the literature an input levy on nutrients has frequently been advocated. However, a uniform input levy on nutrients has substantial allocative disadvantages - such as that least-polluting crops are penalized more heavily than crops from which leaching is considerably greater -, while a differentiation of the rate of the levy invokes substantial agency costs. The most serious objection is, however, that the polluting activities themselves are not taxed, implying incentive distortions.

A levy on nutrient emissions would considerably reduce the chance of incentive distortions. However, to determine the amount of nutrient emissions of a farm, nutrient accountancy is needed, which will generate costs for the individual farmer as well as for the regulating authority. The level and distribution of these costs over the parties involved depends on which institutional arrangement is chosen. The proposal to shift the burden of proof to farmers will reduce the agency costs, but will increase the production costs of farmers.

Although the EU promulgated the Nitrate Directive in 1991, there is essentially no common policy to abate nutrient emissions. According to the subsidiarity principle, regulation on a national level is preferred. However, the instruments used are not allowed to impose arbitrary or hidden trade barriers. As far as we can see the surplus levy, being the variant of a levy on nutrient emissions advocated here, does not generate such barriers, thus fitting in very well with the existing European regulations concerning fair competition as well as with the aim of a single European market.

Despite the highly celebrated subsidiarity principle, the EU has to take policy initiatives in cases where nutrient emissions generate international spillovers. Regarding the increasing eutrophication problems in international waters, such an initiative is urgent. In addition, a European nutrient policy could probably prevent a dramatic shift of intensive livestock farms towards EU Member States employing much lower environmental quality standards.

Notes

[1] The authors thank Floor Brouwer, Nico Hoogervorst, Lex de Savornin Lohman, Herman Vollebergh and Jan de Vries for their valuable and stimulating comments on earlier drafts of this paper.

[2] Animal manure and chemical fertilizers also contain heavy metals such as cadmium, copper, mercury, lead and zinc, originating from concentrated feedstuffs and phosphate ore. The application of manure to land substantially contributes to the accumulation of heavy metals in soils and food. In this paper we leave out this specific environmental problem. We focus on the environmental effects of nutrient leakages due to the application of chemical fertilizers and manure.

[3] According to Merlo (1990, pp. 17-18) 57% of the nitrogen load and 33% of the phosphate load in the Adriatic Sea can be attributed to arable farming and the (intensive) livestock sector in Italy.

[4] In 1989, for instance, 40% of the ammonia depositions in the Netherlands were also emitted in the Netherlands, while this figure for SO_2 was 20% and for NO_x 10% (Bakema *et al.*, 1991, p. 204).

[5] Most of the tapioca and soya is imported from less developed countries (LDCs). These nutrient extractions deplete the soil of large rural areas, implying erosion in the long run. For example, the intensive livestock of the Netherlands is mainly fed tapioca from Thailand and soya from Brazil, which requires an area of arable land that covers at least five times the available acreage of 2 million hectares in the Netherlands (Hoogervorst, 1991, p. 111; IUCN, 1994, p. 59).

[6] Comparative studies concerning the policies for abating nutrient emissions of EU Member States appear to be very rare. The brief overview presented in this section is mainly based on a summary of the current national policies in a study prepared for the Directorate General for Agriculture of the EU (CEC, 1989, Appendix 8) and on partial comparisons and analyses of Conrad (1990; 1992), Baldock and Bennet (1991), Brouwer and Godeschalk (1993) and Rude and Frederiksen (1994).

[7] No nation-wide data were available for Belgium, Italy, Greece and Portugal. From other sources it is known that 40% of the private wells in Flanders exceed the standard of 50 mg nitrate per liter, indicating that the groundwater is seriously contaminated (De Clerq and Sennesael, 1991, p. 132).

[8] Especially the intensively producing farmers in northwestern Europe will be affected by this standard. The mean application rates of fertilizers and manure within the EU show large variations, ranging from, for example, 70-80 kg nitrogen per hectare in Portugal and Spain to 170-450 kg nitrogen per hectare in northwestern France, Belgium, England, Germany, Denmark and the Netherlands (RIVM and RIZA, 1991, pp. 23-26). These figures, however, do not present an accurate picture, because of considerable variations within these regions. In Belgium and the Netherlands, for instance, the application levels for nitrogen easily exceed 500 kg/ha in those areas of these countries where intensive livestock farms are concentrated (De Clerq and Sennesael, 1991; Dietz and Hoogervorst, 1991).

[9] However the so-called 'Third Phase' in the Dutch manure policy (1995-2000) includes plans for the imposition of a regulatory levy on the surplus of phosphates, in combination with a mineral accounting system on farm level (Ministry of Agriculture, 1993).

[10] It is fair to say that the Dutch government is presently in the process of developing regulation of nitrogen application in agriculture. However, definite standards for acceptable nitrate losses are still unknown.

[11] We will return to the problems related to the determination of the maximum application levels in Section 6.

[12] The effects of emissions on the environment are, in general, difficult to predict because of at least three phenomena: synergism, thresholds and delays. In the case of nutrient leaching the effects on nature remain hidden for a long time. For example, it takes decades before nitrogen from manure and chemical fertilizers is washed from the top soil into deeper layers, causing pollution of drinking water resources. Even if nitrogen leakages into the groundwater could be prevented from now on, nitrate pollution of groundwater will still increase considerably for decades into the next century.

[13] The price rise has to be substantial because of the low elasticities of nutrient demands (Hanley, 1990, p. 138).

[14] It is common practice to consider nature's nitrogen processing capacity free to use for farmers. Apparently, the use of nature's nitrogen processing capacity is part and parcel of the bundle of usufruct rights of arable land. Basically, this right seems relatively scarce, implying a willingness to pay for it. For this reason we would expect that the price of land with a large nitrogen processing capacity (such as clay) is higher than that of land with a low nitrogen processing capacity (such as sand), other circumstances being equal. Empirical research on this hypothesis requires at least the

existence of some sort of abatement policy which makes it clear to farmers (as potential buyers or tenants of land) that nature's nitrogen processing capacity is limited.

[15] In the Netherlands the costs of taking and analyzing soil samples currently varies between Dfl. 80.- and Dfl. 100.- per hectare. Rationalization of the procedures and economies of scale will likely reduce these costs.

[16] For this reason nutrient accounts need independent inspection as is the case for financial accounts.

References

Bakema, A.H. *et al.* (1991), 'Verzuring'. In: RIVM, *Nationale Milieuverkenning 2: 1990-2010,* Samsom, Alphen aan den Rijn, pp. 184-213.

Baldock, D. and G. Bennett (eds.) (1991), *Agriculture and the Polluter Pays Principle; A Study of Six EC Countries,* Institute for European Environmental Policy, London/Arnhem.

Baumol, W.J. and W. E. Oates (1971), 'The Use of Standards and Prices for Protection of the Environment', *Swedish Journal of Economics,* Vol. 73, pp. 42-54.

Baumol, W.J. and W. E. Oates (1988), *The Theory of Environmental Policy,* Cambridge UP, Cambridge.

Bennett, G. (1991), 'The Netherlands'. In: D. Baldock and G. Bennett (eds.), *Agriculture and the Polluter Pays Principle; A Study of Six EC Countries,* Institue for European Environmental Policy, London/Arnhem, pp. 87-124.

Bossier, P., F. Huysman, R. Van Stappen en W. Verstraete (1989), 'Mengmest'. In: M. de Coster (ed.), *Milieuzorg in de landbouw,* Pelckmans, Kapellen, pp. 174-205.

Brouwer, F.M. and F.E. Godeschalk (1993), *Pig Production in the EC. Environmental Policy and Competitiveness,* LEI-DLO, publication 1.15, Den Haag.

CEC (Commission of the European Communities) (1989), *Intensive Farming and the Impact on the Environment and the Rural Economy of Restrictions on the Use of Chemical and Animal Fertlizers,* EC, Brussels.

Centraal Bureau voor de Statistiek (1989), *CBS '90; Negentig jaren statistiek in tijdreeksen,* SDU, Den Haag.

Clerq, M. de and K. Sennesael (1991), 'Belgium'. In: D. Baldock and G. Bennett (eds.), *Agriculture and the Polluter Pays Principle; A Study of Six EC Countries,* Institue for European Environmental Policy, London/Arnhem, pp. 125-153.

Comolet, A. and G. Pagnard (1991), 'France'. In D. Baldock and G. Bennett (eds.), *Agriculture and the Polluter Pays Principle; A Study of Six EC Countries,* Institute for European Environmental Policy, London/Arnhem, pp. 154-185.

Conrad, J. (1990), *Nitrate Pollution and Politics; Great Britain, the Federal Republic of Germany and the Netherlands,* Avebury, Aldershot.

Conrad, J. (1992), *Nitratpolitik im internationalen Vergleich,* Edition Sigma, Berlin.

Dietz, F.J. and N.J.P. Hoogervorst (1989), 'Naar een effectief en efficiënt meststoffenbeleid', *Economisch-Statistische Berichten,* Vol. 74, No. 3708, pp. 512-515.

Dietz, F.J. and N.J.P. Hoogervorst (1991), 'Towards a Sustainable and Efficient Use of Manure in Agriculture: The Dutch Case', *Environmental and Resource Economics,* Vol. 1, pp. 313-332.

Dietz, F.J. and K.J.A.M. Termeer (1991), 'Dutch Manure Policy: The Lack of Economic Instruments'. In: D.J. Kraan and R. in 't Veld (eds.), *Environmental Protection: Public or Private Choice,* Kluwer, Boston/Dordrecht, pp. 123-147.

Dubgaard, A. (1986), 'Reconciliation of Agricultural Policy and Environmental Interests in Denmark'. In: M. Merlo, G. Stellin, P. Haron and M. Whitby (eds.), *Multipurpose Agriculture and Forestry,* Kiel, Wissenschaftsverlag Vauk.

Dubgaard, A. (1989), 'Input Levies as a Means of Controlling the Intensities of Nitrogenous Fertiliser and Pestiides'. In: A. Dubgaard and A. Nielsen (eds.), *Economic Aspects of Environmental Regulations in Agriculture,* Kiel, Wissenschaftsverlag Vauk.

Dubgaard, A. (1990), *Danish Policy Measures to Control Agricultural Impacts of the Environment.* Paper prepared for the FAO/ECE Working Party on Agrarian Structure and Farm Rationalization, 12-17 February 1990, Wageningen, Netherlands.

Dubgaard, A. (1991), 'Denmark'. In: D. Baldock and G. Bennett (eds.), *Agriculture and the Polluter Pays Principle; A Study of Six EC Countries,* Institute for European Environmental Policy, London/Arnhem, pp. 32-57.

England, R.A. (1986), 'Reducing the Nitrogen input on Arable Farms', *Journal of Agricultural Economics,* Vol. 37, pp. 13-24.

Erisman, J.W. and G.J. Heij (1991), 'Concentration and Deposition of Acidifying compounds'. In: G.J. Heij and T. Schneider (eds.), *Final Report Second Phase Dutch Priority Programme on Acidification,* RIVM, Report no. 200-09, Bilthoven, pp. 51-96.

Gaag, M.A. van der *et al.,* (1991), 'Fluviale milieuproblemen'. In: RIVM, *Nationale Milieuverkenning 2: 1990-2010,* Samson H.D. Tjeenk Willink, Alphen aan den Rijn, pp. 235-286.

Goossensen, F.R. and P.C. Meeuwissen (eds.) (1990), *Advies van de Commissie Stikstof,* Dienst Landbouwkundig Onderzoek, Wageningen.

Hanley, N. (1990), 'The Economics of Nitrate Pollution', *European Review of Agricultural Economics,* pp. 129-151.

Hoogervorst N.J.P., (1990), 'International Influences on Agricultural Pollution in the Netherlands', *Netherlands Journal of Environmental Sciences,* Vol. 5, 6, pp. 217-224.

Hoogervorst N.J.P., (1991), 'Nutriëntenbeheer en voedselproduktie'. In: RIVM, *Nationale Milieuverkenning 2: 1990-2010,* Samsom, Alphen aan den Rijn, pp. 103-117.

IUCN (the Netherlands Committee) (1994), *The Netherlands and the World Ecology,* IUCN, Amsterdam.

Johnson, S.P. and G. Corcelle (1989), *The Environmental Policy of the European Communities,* Graham and Trotman, London.

Mansholt, S. (1990), 'Nederlandse boeren in een Europees milieu: naar een duurzame landbouw'. In: N.J.M. Nelissen (ed.), *Het milieu: denkbeelden voor de 21ste eeuw,* Commissie Lange Termijn Milieubeleid, Kerckebosch BV, Zeist, pp. 339-357.

Merlo, M. (1990), *Environmental Problems Created by Agriculture in Italy. Legislation and Policies for Improving the Relationship between Agriculture and the Environment.* Paper prepared for the

FAO/ECE Working Party on Agrarian Structure and Farm Rationalization, 12-17 February 1990, Wageningen, Netherlands.

Meyer, B.J.M. en P.F. Lalkens (1988), 'Economische analyse van de bedrijfssystemen op het proefbedrijf OBS'. In: *Themadag Geïntegreerde Bedrijfssystemen,* PAGV, Lelystad.

Meyer, H. von (undated), *The Common Agricultural Policy and the Environment,* WWF International Discussion Papers, No. 1, Gland.

Ministry of Agriculture (Ministerie van Landbouw, Natuurbeheer en Visserij) (1993), *Mest- en ammoniakbeleid derde fase,* SDU, Den Haag.

Neeteson, J.J. (1989), 'Evaluation of the Performance of Three Advisory Methods for Nitrogen Fertilization of Sugar Beet and Potatoes', *Netherlands Journal of Agricultural Science,* Vol. 37, pp. 143-155.

Olsthoorn, C.S.M. (1989), 'Stikstof in de landbouw, waarheen?', *Landbouwkundig Tijdschrift,* Vol. 101, Nr. 12, pp. 22-26.

Pigou, A.C. (1920), *The Economics of Welfare,* edition 1952, MacMillan, London.

RIVM (National Institute of Public Health and Environmental Protection) (1992), *The Environment in Europe: A Global Perspective,* RIVM, Bilthoven.

RIVM (National Institute of Public Health and Environmental Protection) (1993), *Nationale milieuverkenning 3: 1993-2015,* Samson H.D. Tjeenk Willink, Alphen aan den Rijn.

RIVM (National Institute of Public Health and Environmental Protection) and RIZA (Institute for Inland Water Management and Waste Water Treatment) (1991), *Sustainable Use of Groundwater; Problems and Threats in the European Communities,* Report no. 600025001, Bilthoven/Lelystad.

Rude, S. and B.S. Frederiksen (1994), *National and EC Nitrate Policies; Agricultural Aspects for 7 EC Countries,* Statens Jordbrugsoekonomiske Institut, Rapport 77, Kopenhagen.

Schwarzmann, C. and H. von Meyer (1991), 'The Federal Republic of Germany'. In: D. Baldock and G. Bennett (eds.), *Agriculture and the Polluter Pays Principle; A Study of Six EC Countries,* Institute for European Environmental Policy, London/Arnhem, pp. 58-86.

Siebert, H. (1991), 'Europe '92. Decentralizing Environmental Policy in the Single Market', *Environmental and Resource Economics,* Vol. 1, pp. 271-287.

Shortle, J.S. and J.W. Dunn (1986), 'The Relative Efficiency of Agricultural Source Water Pollution Control Policies', *American Journal of Agricultural Economics,* August, pp. 668-677.

Tinker, P.B. (1988), 'Efficiency of Agricultural Industry in Relation to the Environment'. In: J.R. Park (ed.), *Environmental Management in Agriculture; European Perspectives,* Belhaven, London, pp. 7-20.

Willems, W.J. and N.J.P. Hoogervorst (1991), 'Vermesting van bodem en grondwater'. In: RIVM, *Nationale Milieuverkenning 2: 1990-2010,* Samson H.D. Tjeenk Willink, Alphen aan den Rijn, pp. 285-314.

Wit, C.T. de (1988), 'Environmental Impact of the Common Agricultural Policy'. In: F.J. Dietz and W.J.M. Heijman (eds.), *Environmental Policy in a Market Economy,* Pudoc, Wageningen, pp. 190-204.

3 Hybrid Economic Instruments for European Carbon Policy

Paul R. Koutstaal
Herman R.J. Vollebergh
Jan L. de Vries

Introduction

Current logic has it that the problem of climate change is caused by the emission of several so-called greenhouse gases, or GHGs for short, of which carbon dioxide (CO_2) is the most important. CO_2 emissions result from different human activities, but the burning of fossil fuels is by far its major source. The emissions, together with the emissions of other GHGs, are supposed to cause increasing concentrations of GHGs in the atmosphere, which might result in climate change and a rise in the earth's mean temperature. Concern about this scenario has led to different policy proposals to curb those GHG emissions, especially CO_2 emissions. One example of clear commitment is the policy goal set by the European Union to emit the same amount of carbon dioxide in the year 2000 as in 1990. To reach this goal, the Commission has proposed a package of instruments which includes a hybrid energy/carbon tax or excise.

As is well known among economists, under certain assumptions such a tax approach implies a cost-minimizing strategy on the part of the regulatory agency. A tax approach would stimulate carbon reduction at lower cost than would a uniform percentage or absolute level of reduction in Member States, usually the result of more direct forms of regulation. However, the proposal of the Commission turned out to be unacceptable for some of the Member States of the Union. Although the proposal was apparently rejected at the end of 1994, it is now presented as a long-term policy goal for the EU. Member State initiatives to reduce carbon emissions in the meantime fit in this approach very well as a sort of transitional approach. This chapter analyzes to what extent the EU proposal can be transformed in order to improve its political acceptability.

At least four closely related issues relevant for political acceptability are discussed: i) goal attainment or environmental effectiveness; ii) the overall costs of the policy proposal, especially in relation to international competitiveness; iii) target- and burdensharing between countries, given the EU and its Member States as a two-level regulatory structure; and iv) implementation and enforcement problems which are particularly relevant for market-based instruments, such as taxes or tradeable permits (especially if large numbers of polluters and victims are involved). We propose to use selective provision of emission rights to polluters in order to alleviate the high cost of economic instruments for internationally exposed sectors and the distributional problems at the intra-EU level. In such a system, agents no longer have to pay a price for every remaining emission to the

Table 3.1: Global carbon emission from different sources, 1991.

Region	Fossil Fuels		Land-Use Change		Total[1]	
	MMT[5]	Percent	MMT	Percent	MMT	Percent
OECD	2673	39	6	0	2744	40
USA	1334	20	6	0	1352	20
EU12[2]	786	12	n.a.	n.a.	809	12
Japan	285	4	n.a.	n.a.	297	4
Asia	1130	17	251	4	1451	21
Eastern Europe[3]	1200	18	n.a.	0	1230	18
Latin America	259	4	483	7	760	11
Africa	161	2	175	3	358	5
Middle East[4]	263	4	n.a.	0	285	4
World	5686	83	914	13	6827	100

1) Including other sources (flaring, cement production).
2) Germany excluding East Germany; Luxembourg is not available.
3) Including the former states of East Germany (figure 1987), Yugoslavia and Soviet Union.
4) Bahrein, Syrië and Qatar not available.
5) MMT = Million Metric Tons.
Source: World Resources 1990-1991 and 1994-1995.

atmosphere. Compared to the solution chosen by the EU, this proposal leaves the incentives to reduce emissions anywhere at the margin untouched. Furthermore, such explicit grandfathering allows *any* possible solution for target- or burdensharing. Finally, our solution, like the EU proposal, also saves on transaction costs compared to pure economic instruments, even compared to systems of pure grandfathering.

The next section gives a brief overview of 'carbon facts'. These facts show ecological peculiarities such as the fact that climate change is to be seen as a stock externality to which everyone on earth contributes, although there is a wide divergence of carbon emissions among countries and regions (such as the EU). Subsequently, the tax proposal of the Commission is summarized, as well as the criticism to this proposal. To explore whether the current proposal could be improved in order to enhance its prospects for acceptance in the political arena, we return to the economic analysis of instrument choice. First of all, cost efficiency of either pure taxes or auctioned tradeable permits is reaffirmed based on a simple simulation model. Next, we show that these specific economic instruments cause unacceptable policy consequences and we explore to what extent grandfathering of emission rights could contribute to more effective and efficient alternative policy options. We conclude that carefully designed *selective* grandfathering is a case in point and provides better opportunities to be politically acceptable.

Carbon Facts

The greenhouse problem, more specifically the possibility of changing climates with an estimated rise in global mean temperature, is a global phenomenon. Climate change appears likely to be caused by growing concentrations of greenhouse gases (carbon dioxide, CFCs, methane, etc.) in the earth's atmosphere.[1] This is a global problem - both in the sense that its origins lie in the total level of global greenhouse gas emissions regardless of their national origin, and in the sense that individual countries cannot 'cure' their own global warming problems independently of what happens elsewhere in the world. Global warming calls for global policy; the contribution that the EU alone can make to improving the situation is limited as the tables in this section will show.

CO_2 emissions are not evenly spread over all countries and regions. Table 3.1 presents an overall view of sources and amounts of pollutants released, decomposed by regions and time. The world can be divided into more- and less-developed regions, according to their GDP value. The first region consists of OECD countries, such as the USA, Japan, EU-12 (excluding Eastern Germany). Other regions are Eastern European countries, the oil-exporting Middle East, Asia, Africa and Central/South America. Table 3.1 shows that in 1991 fossil fuel-related emissions (mainly found in OECD countries and Eastern Europe) were responsible for 83% of total emissions, while 13% were caused by land-use change, mainly because of the disappearance of tropical rain forests in Latin America, Asia and Africa. An international agreement, which would concentrate only on fossil fuels, would be likely to be one-sided, introducing a carbon reduction policy mainly at the expense of the rich part of the world.

On the other hand, as shown in Table 3.2, if time profiles are taken into consideration richer countries tend to have generated current GDP at the price of much higher accumulated carbon emissions in the atmosphere. Nonetheless, growth rates of carbon emissions between richer and poorer countries show a remarkably divergent trend in the 1970s and 1980s, mainly due to the adoption of more energy efficient technologies in OECD countries and a fast acceleration of deforestation, especially in Indonesia, Thailand and Brazil (not shown in Table 3.2).

Table 3.2: Change in carbon emissions of fossil fuels, 1950-1991.

	Level (MMT)[1)]	Average yearly growth rate			Level (MMT)
	1950	*1950-1965*	*1965-1983*	*1983-1991*	*1991*
OECD	1127	3.0	1.6	1.2	2673
USA	679	2.2	1.3	1.1	1334
EU12	n.a.	n.a.	n.a.	n.a.	786
Japan	27	9.2	5.5	1.6	285
Eastern Europe[2)]	291	6.5	3.6	-0.4	1200
Asia	67	9.5	7.7	2.3	1130
Middle East	4	14.1	9.7	5.6	263
Latin America	36	6.1	6.5	1.0	259
Africa	26	5.1	6.6	0.8	161

[1)] MMT = Million Metric Tons

[2)] Including the former states of East Germany (figure 1987), Yugoslavia and the Soviet Union.

Sources: World Resources 1986, 1990-1991 and 1994-1995.

Correlating the same figures of Table 3.1 with GDP per country changes the picture dramatically, as is shown in Table 3.3. The figures in this table could be interpreted as the amount of carbon emission necessary to generate one dollar of GDP. Poorer countries, in terms of GDP, are now implicitly 'punished' for having a low GDP level, therefore generating relatively high carbon emission coefficients. The choice of figures representing each country's contribution is therefore not unrelated to the policy agreements, although we do not go into this complication in detail (see, for instance, Larsen and Shah, 1994).

As noted before, the problem of climate change is a global stock externality, which implies that individual countries cannot 'cure' their own global warming problems independently of what other countries do. This is even true for the EU as a whole. The emissions of CO_2 caused by the burning of fossil fuels in the EU, for instance, amount to only 12 percent of worldwide fossil fuel-related CO_2 emissions (see Table 3.1). Therefore, curbing emissions only within the EU will have only a small effect on worldwide emissions. Given the growth rates of emissions projected elsewhere in the world, especially in the developing countries, even a large cut in European CO_2 emissions would not be enough to reverse the global upward trend in emissions. Global warming calls for global policy.

On the other hand, if any climate change policy is to be implemented, the developed countries will have to take the lead. The probability that other countries will take the lead is very small because the richer countries are responsible for the greatest amount of emissions both in the past and at present. In line with this reasoning, by signing the Framework Convention on Climate Change (at the UNCED, June 1992), the developed countries (including the transitional economies in Central and Eastern Europe) have committed themselves to stabilizing their emissions of CO_2 and other greenhouse gases at 1990 levels by the end of the decade. This implies that the EU is a relatively useful forum for implementing carbon reduction policies, the more so because the EU is still relatively dependent on oil and gas imports.

Similar problems exist with respect to *intra*-Union differences. Table 3.4 shows the emissions in 1988 and 2010 according to the 'Convential Wisdom Scenario', a reference scenario of the European Commission (excluding Luxembourg, Ireland and the new Member States).[2] This scenario assumes an economic growth in the EU of 2.7% per year in the period 1990-2010 and an oil price of $30 per barrel in 2010. The table shows that Member States with rising emissions according to the reference scenario have to take much tighter measures to stabilize emissions at the 1988 level, than countries with stable or falling emissions. For instance, Greece has to reduce its 2010 emissions by 34% to reach stabilization relative to 1988, while countries like Denmark or the UK will reach this target without any measures at all (or costs).[3] In the next sections we will focus on the question how the EU could reduce its emissions in both an efficient and politically feasible way taking these intra-Union differences into account.

Before turning to this question, we emphasize the issue of climate change is still beset with many uncertainties. Although a broad consensus appears to exist that, on average, some global warming wil take place if emissions of CO_2 and other GHGs continue to rise, the range of natural mechanisms involved remains highly uncertain (for example, whether increased cloud cover could help to counteract the impact on surface temperatures). This complicates risk assessment considerably, especially with respect to uncertainties beyond the range of past experience.

Over and above these scientific uncertainties, there is also scope for legitimate debate about the need for policy measures to combat climate change. As Nordhaus (1991) discusses, measures to reduce GHG emissions would involve significant costs, and these can only be justified if they are exceeded by the costs of uncontrolled global warming. Unfortunately, the costs of a rise in the sea level and of climate-induced changes to agriculture are particularly difficult to estimate. It is even possible that, taking the earth as a whole, the

Table 3.3: Global carbon emissions from different sources (grams carbon per US $ of GDP, 1991).

Region	*Fossil Fuels*	*Land-Use Change*	*Total*[1)]
Eastern Europe[2)]	2136	n.a.	2188
Africa	437	473	969
Middle East	851	n.a.	923
Latin America	233	433	682
Brazil	134	652	797
Other	289	307	616
Asia	217	48	279
China	1780	n.a.	1871
Indonesia	325	776	1179
India	736	23	799
Other	356	222	604
OECD[3)]	158	0	162
USA	238	1	241
EC12	126	n.a.	129
Japan	85	n.a.	89
World	274	44	330

[1)] Including other sources (flaring, cement production).
[2)] Including the former states of Yugoslavia and the Soviet Union and East Germany (figure 1987).
[3)] Excluding East Germany (figure 1987).
Source: World Resources 1994-1995.

effects on agriculture could in fact be in either direction - some regions could gain whilst others lose. It is also conceivable that strategies of adaptation (building sea walls, and moving activities to reflect the change in climate patterns) could prove cheaper than the policy measures required to prevent global warming taking place.

There are thus important areas of uncertainty in relation to both the underlying scientific knowledge and the economics of climate change. Such assessments, however, are far from straightforward as Broome (1992) has illustrated so clearly. We avoid this issue entirely by taking the stated objective, or revealed policy goal mentioned earlier, for reducing CO_2 emissions at the EU level as given.

Current Carbon Policy of the EU

The EU as a whole has committed itself to stabilizing carbon emissions at the 1990 level by the year 2000 (although some Member States have even stricter commitments). As mentioned in the introduction, this commitment was undertaken against the background of international negotiations about concerted action to combat the risk of climate change. To reach this policy goal, the European Commission originally proposed a package of instruments which contained a carbon/energy tax.[4] Since the Commission's proposal was

Table 3.4: European carbondioxide (CO_2) emissions (mln ton), according to reference scenario.

Country	Emissions	
	1988	*2010*
Belgium	109	110
Denmark	61	60
France	374	370
Germany	718	677
Greece	84	127
Italy	399	489
Netherlands	146	165
Portugal	28	51
Spain	196	265
UK	561	505
Total	2676	2819

Source: COHERENCE, 1991.

first released, numerous discussions among Member State authorities have not been sufficient to agree on its implementation. The proposal appeared unacceptable at the end of 1994, but it has been decided to resurrect the concept of the European tax as a long-term policy goal only recently. Member State initiatives to reduce carbon emissions in the meantime are interpreted as a sort of transitional approach.[5]

Although the proposal considers the tax as only one of its carbon policy instruments - other instruments envisaged are, for instance, voluntary agreements with industry on energy and carbon efficiency -, the tax is certainly its most important strategy. For this reason the approach itself, as envisaged in the proposal, gained much support from economists. Economists believe that such an approach is a cost-minimizing strategy from the perspective of the regulatory agency. In theory, taxation promises considerable cost savings compared with uniform emission reduction in the closed economy case. This potential remains even in the open economy case if unilateral action is accompanied by suitable countermeasures such as border controls raising the price of carbon products imported from non-participating countries (Cnossen and Vollebergh, 1992, pp. 31-33; Bovenberg, 1993).

Table 3.5 summarizes the main characteristics of the 1992 EU proposal of the Commission. The tax proposed would have a twofold tax base explaining why it is called a carbon/energy tax. One component of the tax base is related to the carbon content of fossil fuels, while the other is related to the energy content of all non-renewable forms of energy. Non-renewable forms of energy other than fossil fuels (in particular nuclear and large-scale hydropower), however, would still be subject to the energy-related part of the tax, although they would not bear the carbon component. It was envisaged that the two components would be combined in equal proportions, in the sense that half of the tax on a typical barrel of oil would be related to the carbon component and half to the energy component. As proposed, this tax would be introduced in stages, starting at a level equivalent to \$3.00 per barrel of oil in 1993, and then increasing by \$1.00 per barrel annually, until it reached a level of \$10.00 per barrel of oil in the year 2000.

Table 3.5: Main characteristics of the European energy/carbon excise proposal.

Tax aspect	*Characteristics*
Object	Fuels and Electricity: - consumption of fuels (excl. feedstock) - electricity[1)]
Base	Fuels: - 50% energy content - 50% carbon content Electricity: - 100% energy content[2)]
Level[3)]	- ECU 17.70 per ton of oil equivalent or $ 3.00 per barrel of oil in 1993, gradually raised to $ 10.00 per barrel in 2000[4)]
Revenue	- Member States (after subtraction of communitarian monitoring costs) - tax substitution ('neutrality')
Exemptions	- bracketed rates depending on measured costs of energy (excl. VAT) as a percentage of value added - refunds for investment incentives aimed at energy efficiency

[1)] Electricity from solar and wind power is exempted, but from nuclear and hydropower is taxed;
[2)] The carbon component in electricity produced from fossil fuels is taxed through the product tax on fossil fuels;
[3)] This is a *minimum* level;
[4)] Equal to ECU 2.81 per ton CO_2, ECU 0.21 per GJ for fuels, for electricity from hydropower ECU 0.76 per MWh and for other electricity ECU 2.1 per MWh in 1993.
Source: Vollenbergh, 1994, p. 138, based on European Commission, COM(92) 226.

Although the energy/carbon tax had to be introduced through coordinated policy decisions at the EU level, it was to be administered and enforced by the Member States themselves. In addition, the Commission proposal was clear that Member States should use the tax revenues to reduce other taxes, rather than to increase public spending. Moreover, Member States were free to choose which other taxes they would like to reduce, although they would be required to provide the Commission with evidence that this principle of 'revenue neutrality' is fullfilled.

To prevent the European energy/carbon tax from becoming a unilateral action in response to a global environmental problem, the Commission's proposal contains some provisions to prevent undesirable effects on certain industrial sectors. First, there is a 'conditionality clause' that the tax will not be introduced at all unless major competitors take similar measures. Second, individual firms will be exempted from the tax, depending on a case-by-case assessment of the degree of competitive pressure faced from countries not taking equivalent measures, an issue to which we return shortly.

Criticism of the European Carbon Tax Proposal

As noted, the 1992 Commision proposal for the energy/carbon tax gained a lot of support as a cost-efficient policy strategy. Nonetheless, several studies were skeptical on its precise content.[6] Criticisms revolve around the four crucial issues of political acceptability mentioned in the introduction: environmental effectiveness, overall costs in relation to international competitiveness, coordination between Member States within the EU including implementation and enforcement problems.

Environmental Effectiveness

Two issues have been raised concerning the proposed tax base. The choice of the tax base determines to what extent a specific tax proposal reaches its goals, or, in other words, is environmentally effective. First, basing the tax partly on the energy content appears to diminish the efficiency of the tax as an instrument to reduce carbon emissions for any given level of the tax burden. We suppose that the tax was aimed at encouraging reduced use of fossil fuels, especially carbon-intensive fuels, through greater energy efficiency, fuel substitution, and other behavioral responses of energy consumers to the financial incentive provided by the tax. For these responses to operate efficiently, the relative tax rates should be closely related to the relative carbon content of different fuels. In order to achieve any given reduction in carbon emissions the EU energy/carbon tax will have to be set at a higher level than a pure carbon tax, since the energy component reduces the incentive for substitution away from carbon-intensive fuels.[7]

The second issue concerning environmental effectiveness refers to the choice of a 'final' type of energy/carbon tax (Smith and Vollebergh, 1993, pp. 213-215). According to the current proposal fuels are taxed where they are sold to final fuel consumers. However, most fuel products go through various stages of production, processing or refining. Unlike a 'primary' tax, which would be levied at an early stage in this chain, viz. on primary fuels where they are mined or extracted, a 'final' energy/carbon tax is based on carbon and energy content of those final fuel products and therefore does not reflect energy losses and carbon emissions during processing. In principle, it would be possible to adjust the tax rates for different fuels, so that the tax rates reflect their current energy and carbon content *plus* the losses and emissions during processing. It is clear from the Commission's proposals, however, that such adjustments were not envisaged. The tax thus provides no incentive for fuel processors to reduce emissions during processing and is liable to distort the energy market towards highly processed forms of fuel.

To illustrate, refineries which transform crude oil into mineral oils can (depending on the process used) use large amounts of energy in the course of refining. The final refined fuel products contain less carbon and energy than the original primary fuels. The energy used in refining leads to carbon emissions, which should be taxed on an equal basis with the carbon embodied in final fuels, otherwise a subsidy is given away to the refineries. If the use of energy and carbon emissions during the refining process are disregarded in the tax rate on final products, an undesirable incentive is given towards the use of highly refined fuel products, in which as much carbon emission as possible has taken place before the tax is applied.

The way in which the energy/carbon tax would be applied to electricity, however, is more in conformity with an efficient incentive structure, although not entirely. The carbon component on electricity would be levied on a 'primary' basis; the tax would be paid on the carbon contained in the input fuels, not on the (zero) carbon content of the output fuel. However, this logical approach has not been carried over to the treatment of the energy component of the tax. Here the energy content of the output would be taxed instead of the inputs, and at a rate higher than other energy, presumably to reflect average (but not actual) energy losses during generation.

Problems with Overall Costs and International Coordination

The second set of criticism relates to the proposed international coordination of the tax, which determines to what extent industries within the EU face a decline in their international competitiveness. International coordination is clearly the key to effective control of GHG emissions as we saw before. There is little point in unilateral EU action, given the limited impact on the global problem at stake. At the same time, however, the 'conditionality clause' introduced in the proposal is so restrictive as to inhibit progress on the measure altogether. The clause leaves little room for international leadership and requires an unrealistic, and unnecessary degree of international uniformity on the precise form of measures taken within different countries.

The Commission has made it clear that the tax would only be introduced if 'a similar tax or measures having an equivalent financial effect' are introduced by other OECD countries. Requiring measures with 'equivalent' financial effect' could be interpreted so as to rule out the possibility that other countries could choose a different package of measures based on regulation or other 'conventional' instruments. It is far from clear that there is a need for the EU to set requirements for the form which other countries' policy interventions should take. This goes well beyond what is required by international policy coordination.

In addition to the 'conditionality clause' the Commission proposed that special provisions granting temporary tax exemptions, tax reductions and refunds should apply to 'firms with a high energy consumption that are seriously disadvantaged on account of an increase in imports from third countries' which had not taken similar measures. In these cases Member States could grant firms a graduated reduction in the energy/carbon tax payable, if energy costs amount to at least 8 per cent of value added. Furthermore, Member States could refund tax revenues in the form of subsidies for investments which improve the efficient use of energy or which limit carbon dioxide emissions directly.

These rules for tax exemptions, reductions and refunds are rather complicated and would provide a big incentive for rent-seeking behavior. They would also severely complicate administration of the tax increasing administrative costs and the need for enforcement resources. Given that entitlement to rebate depends on energy costs above a given threshold share of value added, a clear incentive exists for the foundation of subsidiary companies producing the bulk of the energy consumption. Energy cost as a percentage of value added in the subsidiary company will easily become high enough to justify claims for tax reduction. New firms will come into existence, simply in order to minimize tax payment. A further concern, as discussed below, is that the rules for exemption may become open to abuse. Given that exemptions are discretionary and negotiable, there is a risk that their application may be influenced by considerations not strictly within the original objectives of the rebate provisions, and may become a route by which Member States provide a degree of 'backdoor' subsidy to either vulnerable or favored industries. Another drawback of the proposed exemptions and reductions is that the incentive to reduce energy use and carbon emissions would be weakened or removed, while the refunding of tax proceeds in the form of (100%) subsidization could create distortions in the pattern of investments in energy conservation and emission reduction.

Intra-Union Coordination and Enforcement Problems

The final set of issues relate to the coordination between Member States or *intra-Union coordination*. According to the principle of subsidiarity as codified in the Maastricht Treaty (article 3b), the EU should take action "if and in so far as the objectives of the proposed action cannot be sufficiently achieved by the Member States and can therefore, by reason of the scale or effects of the proposed action, be better achieved by the Union". It is also stipulated that "any action by the Union shall not go beyond what is necessary to achieve the objectives of this Treaty". The subsidiarity principle is often used to dispute the adequacy of environmental policy measures by the EU, and as an argument to leave at least the choice of policy instruments at the level of Member States (Brouwer, 1994). Therefore, it might be argued that the introduction of this first EU-wide tax is contrary to this subsidiarity principle. In fact, this fundamental objection, strongly emphasized by the UK, was ackowledged in the final decision to reject the Commission proposal and to give priority to Member State initiatives.

A further problem regards the issue of target- and burdensharing between Member States. In spite of efficiency gains for the EU as a whole, the energy/carbon tax may well lead to disadvantages for some countries in terms of higher abatement costs, compared with the costs of a uniform reduction target for some individual Member States.[8] As is clear from Table 3.4, any measure curbing CO_2 emissions would hit countries with lower GDP levels hardest at the outset. Therefore it is hardly surprising that the proposal for the energy/carbon tax encountered considerable resistance from some (Mediterranean) countries.

The judgement is relatively positive as far as the proposals' consequences with respect to the transaction costs of implementation are concerned (Vollebergh, 1994b). By taking fossil fuel-based carbon and energy as principal tax bases, the proposal takes advantage of the currently existing tax administration as much as possible. Only some additional tax enforcement will be necessary. For only some fuels, in particular natural gas, LPG and solid fuels, administrative procedures in the Member States are lacking (Hoornaert, 1992, p. 87). Nonetheless, these procedures are essential for a comprehensive approach to all fossil fuels which is necessary for the intended intrafuel substitution of carbon extensive for carbon intensive fuels.

Unfortunately, considerable doubts have been raised concerning monitoring and enforcement of the tax. The proposal of the Commission itself is rather silent on this important issue. By referring to the principle of subsidiarity, the Commission seems to delegate all monitoring and enforcement issues to the Member States themselves. Questions arise, however, to what extent Member States can be expected to comply with the requirements of the system. As is the case with any other important international agreement, some monitoring and enforcement has to be centralized. Monitoring of compliance is especially important in view of the significant incentives for free rides: considerable parts of the benefits of the tax are global, while non-compliance by an individual country will have little impact on the extent to which they experience these benefits. On the other hand, substantial adjustment costs in certain industries would arise from the tax, which might be avoided if the tax is not rigorously enforced. Without some monitoring and enforcement, there is a risk that Member States will be prepared to be unduly lax in the application of the tax to national industries, for example, through generous use of the scope for exemptions or refunds. This could provide a channel by which the operation of the tax introduces distortions into the pattern of industrial competition within the Union.

Cost Efficiency of Pure Taxes or Tradeable Permits

As mentioned earlier, the 1992 Commission proposal to introduce a European energy/carbon tax has been swept from the negotiating table, in any case for the foreseeable future. However, the EU policy target itself, to stabilize carbon emissions at the level of 1990 by the year 2000, still exists. Besides, many policy makers seem to be committed to the idea that taxation is a vital element of any instrument mix to reach this target. We conclude therefore that either other, politically acceptable, carbon taxes or alternative policy instruments must be found, or the target itself will turn out to be unrealistic. Before turning to the question if and how the EU tax proposal could be improved, we return to the economic analysis of instrument choice. This section, first of all, provides further evidence on the potential efficiency gains of a tax or tradeable permit approach compared to uniform regulation based on a simple simulation approach.[9] The simulation approach also shows which problems in terms of political acceptability are raised if cost-efficient instruments such as taxation or auctioned tradeable permit are used to reach the policy goal. We include the analysis of tradeable carbon permits even though they are not yet considered in the political debate. Auctioned tradeable permits, however, provide a serious cost-efficient alternative in the European context as well.

In theory a regulatory agency could choose between two cost-minimizing regulatory strategies given a predetermined carbon reduction policy. Any given vector of final outputs along with a set of specified standards could be achieved with the use of specific taxes or auctioned tradeable permits inducing necessary adaptation of production processes at minimum cost to society (Baumol and Oates, 1988). Elsewhere we simulated this theoretical result for a carbon stabilization goal in 2010 for ten Member States of the EU based on estimates of country-specific abatement costs (Koutstaal, Vollebergh and De Vries, 1994). Table 3.6 reproduces the results of the two cases estimated: (1) a uniform reduction of emissions in 2010 by 13% or 366 million tons of CO_2; and (2) a tax of 20 ECU per ton CO_2 or a system of auctioned tradeable permits reducing emissions in 2010 by almost the same amount or 365 mln. tons of CO_2 (implying a price for tradeable permits of 20 ECU per ton CO_2).

As predicted by the cost minimization theorem, Table 3.6 shows that the overall abatement costs would be lower if instruments like a tax or auctioned tradeable permits are used compared with a uniform (percentage) reduction scheme. According to these calculations the total cost advantage would amount to 27%. This result illustrates that, from an aggregate cost-efficiency point of view, a choice in favor of a tax or permit system compared with a uniform reduction percentage is obvious. Thus potential efficiency gains for the Member States of the EU exist in the case of a CO_2-reduction policy based on carbon taxes or auctioned tradeable carbon permits applied across the board. Such economic instruments are called *pure* economic instruments because they reflect the tax base as intended by the regulatory goal, with no exemption for any emitter at all.

Unfortunately, political acceptability of pure economic instruments will be even worse compared to the policy proposed by the EU. A policy based on pure taxes or auctioned tradeable permits in an open economy with the EU emitting only 12% of global emissions, is useful only if other countries were to act now as well or follow before long. The assumption underlying the simulations that the EU acts as part of an international agreement, is obviously far from reality. Lack of international coordination would imply unilateral action on the part of the EU. Polluters would not only face a loss of market share due to higher costs reflecting the abatement costs and the extra payments on taxes or permits, but also due to the fact that competitors from abroad do not face a comparable rise in costs. Moreover,

Table 3.6 Abatement costs of different carbon abatement policies in the EU.

Country	Uniform reduction target		Carbon tax or tradeable carbon permits		
	abatement cost	abatement cost per capita	abetement cost per capita	abetement cost	tax or permit revenue
	mln. ECU	*ECU/capita*	*mln. ECU*	*ECU/capita*	*mln. ECU*
Belgium	467	47	104	11	2055
Denmark	34	7	104	20	929
France	672	12	344	6	6708
Germany	1358	22	881	14	12142
Greece	73	7	374	37	1792
Italy	494	9	818	14	8144
Netherlands	369	25	141	10	3019
Portugal	29	3	118	11	760
Spain	155	4	624	16	3917
UK	1495	26	243	4	9613
Total	5147	16	3753	12	49078

Source: Koutstaal, Vollebergh, and De Vries (1994).

such a unilateral policy might not only cause a loss of competitiveness and income, but also have the adverse effect of a global *rise* in carbon emissions because of a relocation of carbon-intensive industries to non-regulated areas (Hoel, 1991).

Our simulations also show that large amounts of tax or permit revenues are involved in order to obtain only a relatively small amount of reduction. With price elasticities in the normal range, total costs of both pure taxes and auctioned permits are a multiple of the abatement costs. The public revenue estimates presented in the last column of Table 3.6 show those extra costs if a pure tax or auctioned permits are used. With marginal costs of carbon abatement of 20 ECU per ton of CO_2, the aggregate abatement costs are less than 8% of the amount of revenue generated. This effect enhances the problems related to unilateral action already mentioned.

Of course, it can be argued that the revenue raised by a tax or auctioned permits is not a loss for the country implementing such a tax reform (Smith, 1992). If revenue neutrality is pursued, the total tax burden might be kept constant for the regulatory agency could, for example, lower other distortionary taxes.[10] This certainly alleviates unfavorable effects on the competitiveness of certain industries and even some (currently non-existing) industries might even gain a competitive advantage due to lower other marginal taxes (e.g. labor taxes) compared with the status quo. However, especially the carbon-based energy-intensive and exposed industries will not be compensated entirely by such indirect revenues as lower taxes on labor. Without countermeasures relocation of these industries outside the EU remains attractive.

Furthermore, the simulations illustrate which *intra*-Union differences across Member States cause further problems for political acceptability. Total costs and abatement costs per capita differ considerably between the different policy options. In the tax or permit scenario, Denmark, Greece, Italy, Portugal and Spain face higher abatement costs, while

Belgium, France, Germany, the Netherlands and the UK face lower abatement costs. Also the level of costs per capita in the various countries is substantially different. In the efficient solution, costs per capita are above average in Denmark, Germany, Italy, Spain and especially Greece. Abatement costs could rise up to 21% of the tax payments or permit costs for individual countries, although these figures are partial equilibrium calculations taking neither redistribution of the generated revenue nor concessions in the incentive systems into account.[11]

Finally, our simulations, like the cost efficiency theorems of economists in general, are based on the assumption that monitoring and enforcement of taxes and auctioned permits is costless. This assumption is unsatisfactory, since any comparison of instruments should take these costs explicitly into account (see Chapter 1). At first sight, a pure tax system might avoid the drawback of high transaction costs. As the discussion of the EU proposal already clarifies, if the tax could be based on inputs rather than on emissions, and carbon use is already taxed in the status quo, transaction costs of pure taxes need not be high (Vollebergh, 1994a).In contrast, however, transaction costs seem rather high in a pure system of auctioned tradeable permits. This instrument presupposes that all polluters should be monitored and could take part in the permit market. Particularly the costs of trading in carbon permits are probably large for households. Moreover, registration of all individual transactions of consumers drives up administration costs for the regulatory agent. Therefore, transaction costs are rather high in this system if literally millions of polluters participate.

Clearly, our simple simulation illustrates why political acceptability of pure taxes or pure auctioned tradeable permits is low. Although cost efficiency is guaranteed due to the taxation of every marginal emission, this result is obtained at the expense of the international competitiveness of the exposed industries and the growth potential of the Southern Member States (as long as the EU acts alone). Concerning the transaction costs of pure systems taxation does a better job than pure auctioned permits, which might explain the choice for a European tax rather than an EU-wide tradeable permit system. Now, the obvious question is whether economic theory can be helpful in finding other solutions which are more politically acceptable without losing too much of their economic attractiveness.

Pure Economic Instruments and Political Acceptability

We are after incentive mechanisms which might perform better on the four dimensions of political acceptability. The most obvious solution is to reduce overall costs for polluters without affecting the incentive to abate emissions in the most cost-efficient way. Economic theory suggests that high overall costs of pure systems can be alleviated by separating efficiency and equity at the margin through *grandfather rules* (Pezzey, 1992; Zodrow, 1992). By granting free carbon emission rights to countries, industries or firms, total costs of a particular economic instrument become lower, while the incentive mechanism to reduce emissions at the margin remains. The costs due to the financing of government expenses is released because agents no longer have to pay for their remaining emissions. The application of grandfather rules with regard to tradeable permits is well known,[12] but in theory environmental taxes based on generalized grandfathering might be considered as well.

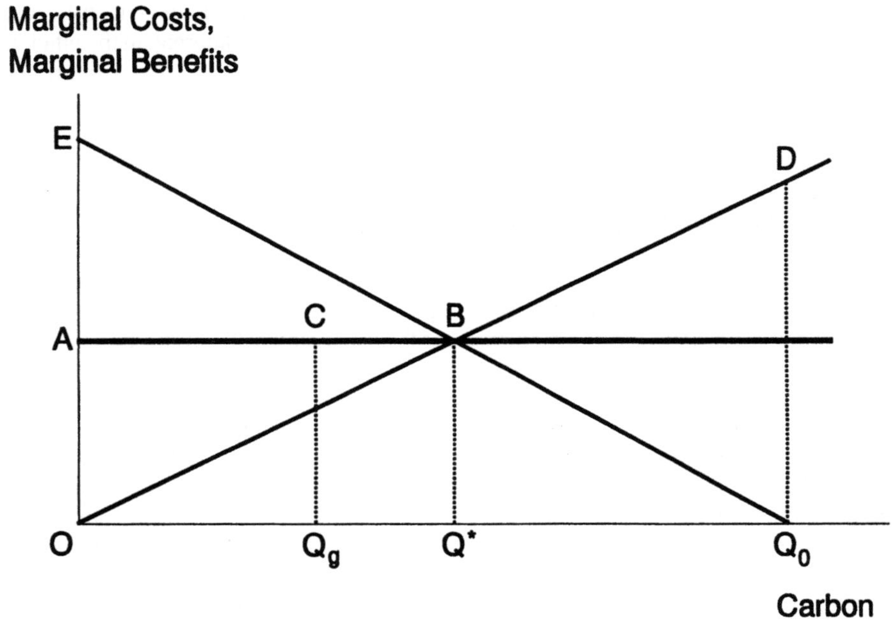

Figure 3.1: Optimal pollution and choice of instruments.

We illustrate the essence of this extended set of 'pure' systems with the familiar linear Figure 3.1. One carbon permit is assumed to be the equivalent of 1 ton of carbon implying that one carbon permit allows the use of a quantity of fossil fuels containing 1 ton of carbon. Ideally, in all cases a similar market price signals the same marginal value to the individual agent. A tax with a carbon base set at the level Q^*B induces agents with total *de facto* carbon emission rights Q_0 in the status quo to abate carbon emissions. Given the (average) marginal abatement cost curve Q_0BE, agents reduce carbon emissions until their marginal costs of further abatement equal the unit tax. After the reallocation, total costs comprise both the tax payment $OABQ^*$ plus the abatement costs triangle Q^*BQ_0. A pure system of auctioned tradeable permits gives exactly the same results, with the agents buying the amount OQ^* at price Q^*B at the auction. As long as the revenues of the auctioned permits or the tax are not earmarked but used for public expenditures instead, current polluters face a higher burden of public funds than in the status quo. Therefore, the same resistance to auctioned permits *and* a carbon tax is to be expected.

In the case of pure grandfathering of tradeable permits, all permits necessary to reach the stated reduction goal are distributed for free to current polluters (for an amount of OQ^* in terms of Figure 3.1). This restricts *de facto* pollution rights compared with the status quo (Pezzey, 1988). A polluter preferring to sustain his original emission level Q_0 is obliged to buy permits from other polluters. This drives up the permit price. In turn, the higher permit price creates the incentive to minimize costs and emissions, until emission level Q^* and permit price Q^*B are reached. These results are identical with those of a pure tax or auctioned permit system, except that in this case the polluter is only faced with his abatement costs Q^*BQ_0. In this way, the overall costs for the polluters are considerably reduced and negative effects on their competitiveness are mitigated substantially.[13]

In theory, a similar pure tax system in which the polluter is only charged for his carbon emissions beyond Q^* is imaginable. Conceptually, a *marginal tax* only taxes emissions which can be abated at marginal costs lower than or equal to the tax rate, or, in the case of carbon, the *infra*marginal use of carbon-containing fossil fuels. The marginal tax leaves the incentive for emission reduction intact and therefore minimizes abatement costs. Note, however, that the allocation of tax exemptions is different compared to the grandfathering of tradeable permits. Because a tax exemption as such is not tradeable, no polluter should receive a tax exemption exceeding his marginal use of carbon. Otherwise, he would not reduce emissions up to the point where his marginal abatement costs equal the tax rate, thus disturbing cost minimization.

Apart from the fact that systems of pure grandfathering leave the government without any revenue at the margin, thus destroying any possibility for tax reform, they have several other drawbacks. Both options for purely grandfathered systems cause serious implementation problems. For the marginal tax system this is hardly surprising. A 'zero-revenue tax' presupposes either a tax authority knowing *ex ante* the intersection between abatement costs and tax level for each individual polluter (point B in Figure 3.1), or the retribution of tax revenues to individual taxpayers, without weakening the incentive to reduce his emissions to Q^*. Moreover, from a more dynamic perspective, a marginal tax has the disadvantage that any incentive to reduce emissions below Q^* disappears. A pure system of grandfathered tradeable permits does not have the latter drawback, but presupposes that all polluters who receive permits for free should be monitored and could take part in the permit market. Therefore, transaction costs are still high in this system.

We conclude from these observations that pure systems, whether fully grandfathered or not grandfathered at all, are capable of realizing the carbon reduction goal in an efficient way. Although purely grandfathered incentive systems solve acceptability problems due to the high overall cost burdens of pure economic instruments, they either create (marginal tax system) or do not solve transaction cost burdens (tradeable permits). Thus all pure systems create serious acceptability problems due to either high overall cost burdens or implementation problems or both. Political acceptability of all four pure incentive mechanisms is therefore low, and the question remains whether mechanisms exist which can solve these problems simultaneously in a satisfying way. The next section shows that *selective* grandfathering is the solution, for incentive mechanisms based on this principle take stock of both the high overall cost burden and the implementation costs. Furthermore, flexibility with respect to the choice of grandfather rules in such systems also allows issues related to target or burdensharing between Member States within the EU to be settled.

Alternative Policy Design: Hybrid Carbon Incentive Mechanisms

The preceding section showed how grandfathering of emission rights might alleviate high overall cost burdens for polluters due to regulation by pure carbon taxes or auctioned tradeable permits. Unfortunately grandfathering is costly if applied across the board. This section shows how *hybrid* carbon incentive mechanisms, based on *selective* grandfather rules, might solve both problems at the same time. We show that carefully chosen selective grandfather rules optimize the advantages of grandfathering together with the cost of enforcement.[14]

Under selective grandfathering rules only some of the available emission rights (corresponding with the emission target) are given away to the current polluters. Those rights are used to reach certain equity goals of the regulating agency, for instance objectives related to target or burdensharing. In the theory of optimal tax reform such rules are well established as instruments to enable Pareto-improving tax reforms (Zodrow, 1992). Their focus is on prevention or at least alleviation of individual welfare losses caused by tax reform measures, despite the existence of welfare gains on the aggregate, social level.

Obviously, selective grandfathering is possible with taxes, tradeable permits or combinations of taxes and tradeable permits. Subsequently we briefly describe how these systems work and perform.

Carbon Taxes with Thresholds

In the framework of tax reform, selective grandfather rules are usually implemented by exempting certain tax payers entirely or beyond a certain level from their tax liability (e.g. the provisions proposed in the EU carbon/energy tax). Quite a different design, which is particularly interesting in our context, is the granting of tax thresholds to each tax liable agent separately.[15] Figure 3.1 illustrates that the polluter is given *de facto* emission rights for free, e.g. for the amount of OQ_g. Thus government revenues will be reduced by the amount $OACQ_g$, and, by implication, overall costs of the polluter are reduced by the same amount. So, the granting of tax thresholds results in alleviation of the overall cost burden without affecting the incentive to reduce emissions at the margin for individual agents, provided that the magnitude of the threshold is smaller than their marginal level of emissions. Thus a hybrid tax system based on tax thresholds differs from a pure carbon tax in that (part of) the polluters receive tax exemptions to some extent. Only the emissions exceeding the threshold are taxed, while the emissions below this level are exempted.[16] Therefore, such a hybrid system offers an opportunity for fine tuning of distributional effects.

Note that tax thresholds are a non-tradeable emission right, implying that a consumer or producer cannot benefit directly from trade even if opportunities to reduce emissions exist below the exempted emission level. In this case no market exists to capitalize on. Because the thresholds themselves are non-tradeable, no agent should receive a emission right exceeding his marginal demand for carbon. Otherwise, the agent would not reduce emissions up to the point where his marginal abatement costs equal the tax rate, preventing abatement costs being minimized.

This problem seems to complicate matters, for demands on the regulating authority rise considerably with such a system. In principle, the authorities have to know the abatement cost function of each polluter in order to be sure that the tax threshold does not exceed marginal use. Thus we run into a paradox, for one of the alleged advantages of using economic instruments is that the authorities would not need such information in order to be efficient (Baumol and Oates, 1988, pp. 160-163). This paradox is solved by allocating tax thresholds below an uncertainty margin of marginal use. In that case the authorities only need to know this margin and a crude estimate of its associated abatement costs. This information is often available from statistical or engineering bureaus, and tax thresholds could be assigned accordingly to specific industries. Nevertheless the complexities of allocating tax thresholds should not be underestimated. Problems include the demarcation of industries and the choice of an appropriate method for determining the tax thresholds (SER, 1993, p. 71). In cases where allocation of thresholds is too costly, the regulatory authority could simply apply full tax exemptions instead of using the tax with threshold system.

Who should receive emission rights for free? Tax thresholds offer a relatively simple procedure for alleviating overall costs for particular internationally exposed industries, but also for burdensharing between (intra-EU) and within Member States. Solutions to this problem are open to negotiating for any solution is possible as long as the thresholds do not exceed marginal use of carbon by countries, industries or households. Therefore, such a carbon tax need not be harmonized with similar thresholds all over the EU. Credits can be allocated by each Member State in accordance with the country's distributional preferences, which also leaves substantial room for considerations regarding subsidiarity.

In principle, not much difference exists between industrial consumers of carbon and households. Administrating tax thresholds for each household separately, however, might easily become too costly, due to the large number of agents concerned. As far as energy sources are distributed via centralized technical infrastructures (e.g. natural gas and electricity), interesting opportunities exist for saving on transaction costs. In such cases supply companies can be enlisted to tax only consumption which exceeds the exempted level.[17] Effective monitoring is much more difficult if households also use other energy sources such as coal, for which administration at the household level is much more difficult to enforce. For example, household coal in the UK is available from gas stations involving high costs of monitoring. Therefore it seems preferable to leave out tax thresholds for consumers and to charge them a full tax instead.

Implementation of the tax with threshold system is rather simple, and could be left entirely to the Member States themselves. Every firm in a particular Member State receives vouchers for the amount of its tax thresholds allowed. In turn, the firm hands over these vouchers to their suppliers of fossil fuels. The suppliers only charge tax for the fossil fuels not covered by the vouchers. Suppliers pass on the collected tax revenue ánd the vouchers to the authorities in the Member State where they are located. Thus monitoring and enforcement is easily combined with the existing institutions for levying excises on fossil fuels as far as they already exist. Only a limited number of suppliers have to be checked: producers and importers of fossil fuels in a particular Member State. In the Netherlands, for example, only 40 to 50 of such firms exist, and a high level of compliance is easy to organize. In order to test compliance in individual Member States, the EU only would have to monitor each country on a regular basis.

Hybrid Tradeable Carbon Permits

In the framework of tradeable permit systems, selective grandfathering is simply implemented by giving *de jure* tradeable permits for free to specific polluters. Thus, in a hybrid system of tradeable carbon permits, some of the permits are grandfathered and others are auctioned by the regulatory agency. Polluters who receive permits for free, only have to bear the abatement costs provided that the total number of grandfathered and auctioned permits is in accordance with the intended lower overall emission level. The exact amount of granted permits in such a scheme is unimportant, and any amount of granted emission rights could be chosen.

Again, the obvious problem is to determine who will receive permits for free and who will have to buy auctioned permits. Whatever the exact boundary between those who receive permits for free and those who have to buy them initially, the regulatory agency has to differentiate between both categories. In the system we have in mind, the grandfathered permits are provided directly to the individual members of the exposed industries, while the auctioned permits are sold to the suppliers of fossil fuels. The suppliers of fossil fuels subsequently raise their prices with the costs of the auctioned permits signaling the scarcity price to all consumers. Polluters who received permits for free may turn in their permits and buy fuel for its original price. Auctioning and grandfathering are thus used side by side in this system (compare Figure 3.2).

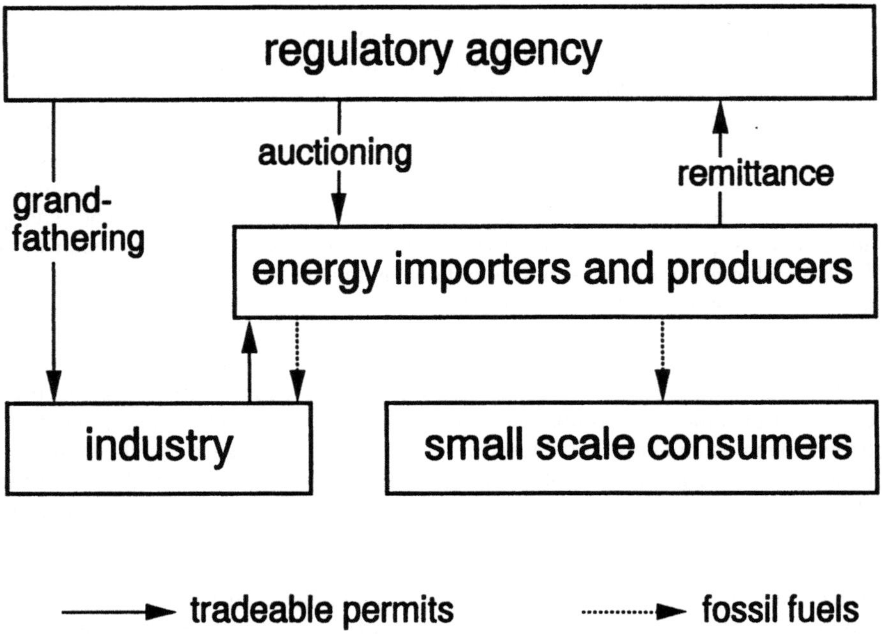

Figure 3.2: A hybrid system of tradeable carbon permits.

Permits for the fossil fuel sold to the households should not grandfathered to the suppliers of fossil fuels in the system envisaged here. In this case the suppliers would earn a rent receiving permit proceeds from the fuels they sell. Furthermore, auctioning of permits to suppliers of fossil fuels does not cause any damage in the international market under the permit scheme described here: every supplier, whatever his country of origin, has to acquire permits in order to sell fossil fuels within the EU. Thus, all suppliers face equal terms of trade on markets within the EU, but also on world markets as long as the producers are exempted for their exports (assuming a competitive market in fossil fuels).

Another problem is the condition on which polluters receive their allotment of permits. This condition might be based on actual emissions in a reference period, corrected for the intended emission reduction. Thus, polluters who reduced emissions prior to the reference year are 'punished' if they receive fewer permits than polluters who did not reduce pollution. A solution to this particular problem is to put a limit on the number of permits for less energy-efficient polluters. Unfortunately, this requires information on the relative efficiency of energy users. Again cost of information should be counted against, in this case, fairness of the system. Such drawbacks have to be accepted, as is usual in tradeable emission schemes which have been put into force until now (and, for example, in the milk quota scheme of the EU!).

Within a European system of tradeable carbon permits, permits can be allocated in any way among the different countries.[18] Thus, cost burdens of CO_2-abatement across the Member States of the EU could be manipulated in accordance with any distributional preference. For example, Southern States with lower income per capita, might receive relatively more permits than Northern States. Even if the permits are sold at a central European level, revenue could be divided according to a pre-arranged key (as in the VAT Clearing House system). Note also that the authorities can issue permits every year, adjusting the number in order to reduce CO_2 emissions by a specific amount. Such a system leaves room for any policy change, for example, if new knowledge is acquired about climate change.

The equilibrium outcome in the case of tradeable permits is reached by trade in permits within *and* between countries. Transfers of financial resources from countries where permits are relatively scarce to countries where they are relatively abundant is inevitable. The amount of transfers depends on the initial distribution of permits ánd on the availability of low cost options for emission reduction. In other words, costs for certain Member State are not necessarily borne by that State allowing any tradeable permit schemes to function as a burdensharing mechanism. Therefore, secondary trade between owners of permits should be allowed to take place within ánd between Member States for this establishes the efficiency gain. Moreover, reallocation of cost burdens among polluters in the EU has to be accepted with this regulatory mechanism.

In a European system of carbon permits, monitoring and enforcement need not be centralized and individual Member States could organize local auctions and markets. Nevertheless, monitoring the performance of Member States is necessary like in the tax system. Following the analysis of Heister et.al. (1992), implementation of a tradeable permit scheme cannot be delegated completely to the Member States under current EU rule. At least, the total number of permits and the initial distribution between Member States must be decided upon at the Union level.

In the system envisaged here, control and enforcement is no more difficult to organize than in the tax with threshold system. Suppliers of fossil fuels are obliged to hand over carbon permits (e.g. once a year) to the authorities for the carbon contained in the fossil fuels they sold on the market ('remittance' in Figure 3.2). They acquire permits either by buying them at the auction or from the industrial firms that have received them for free. Control can be combined with existing institutions for levying excises on fossil fuels as well and a similar limited number of suppliers have to be checked as with the tax with threshold system. Again a high level of compliance seems to be possible, rendering this hybrid tradeable permit system just as feasible as a carbon excise.

Toward a More Effective and Efficient EU Carbon Policy

Our search for more acceptable economic instruments has finally come to an end. In view of the four criteria for political acceptability *selective* grandfather rules provide interesting opportunities for politicians. The upshot of these systems of selective grandfathering is that several emitters get lump-sum transfers in the form of either tradeable or non-tradeable emission rights through the design of the incentive mechanism itself. Polluters are protected against high overall cost burdens of pure taxes or auctioned tradeable permit systems. In each case the political authority is free to choose the amount of carbon emission rights to be grandfathered. Thus, the political authority can separate efficiency and equity *within* the regulatory system itself, while some room for tax reform remains due to the revenue raised by the emission rights not provided for free to polluters. Thus, selective grandfathering allows the regulating authority to solve problems of international coordination *and* problems of control and enforcement costs of other solutions.

As the section on the European carbon/energy tax showed, selective grandfather rules are in fact included in the 1992 Commission proposal for an EU energy/carbon tax. The provisions for energy-intensive industries solve part of the problem of high overall costs burdens for these (usually internationally exposed) industries. Unfortunately, these allowances weaken the incentive to reduce emissions at the margin forming a serious threat for economic efficiency. Another disadvantage of the tax proposed by the EU is its insensitivity to distributional considerations, largely due to its uniform application everywhere in the EU. Moreover, as observed, the economic efficiency of this system might be at risk due to the high transaction costs involved. Thus, the particular design of the tax chosen by the EU is far from optimal, even though it takes advantage of selective grandfathering.

In contrast, our systems of selective grandfathering provide a much better solution if issues of political acceptability are taken into account. The hybrid carbon incentive mechanisms described in this chapter leave the incentive to reduce emissions everywhere at the margin untouched. Consequently, overall carbon abatement in the EU will be more efficient. Furthermore, our incentive schemes allow for distributional considerations in two ways. First, tax thresholds or grandfathered tradeable permits provide a concession for high overall costs for polluters, although incentives to relocate to countries outside the EU (if they do not follow EU policy) are not completely removed. Some rise in the cost for current polluters cannot be avoided, but this is justified by the expected benefits. Second, Member States might prefer quite different allocation mechanisms for either the tax thresholds or the grandfathered permits. In other words, the hybrid tax or permit systems need not be harmonized across the EU leaving much more room for considerations based on 'subsidiarity'.

As far as practicality of our proposal is concerned, the tax with threshold system in particular is no more difficult to implement than the EU proposal. Although some trade-off remains with respect to who exactly gets a tax threshold, the system is rather easily enforced if advantage is taken of already existing administrations and statistical bureaus. Like the EU proposal, additional investment in monitoring and enforcement is necessary only for some fuels in order to reach the most comprehensive tax base. Comprehensiveness of the overall carbon tax is fundamental if the tax is to obtain its environmental effectiveness. The tradeable permit system might look more attractive if one realizes that it is easily combined with a tax system. For instance, a tradeable permit system with grandfathered emission rights to industry, might be used alongside with a tax on the consumption of fossil fuels. This allows the EU to experiment with a tradeable permit system in a precisely defined area.

Finally, we would like to warn against the trend of contrasting subsidiarity and coordination. As the carbon case study reveals, a prima facie argument exists for the EU to act on behalf of the Member States if a particular environmental issue has a global dimension. Unfortunately, coordinated action is too often put on a par with inflexible harmonization. The current proposal for a European carbon/energy tax is a case in point. Member States can only choose to spend the public revenue raised by the tax, and merely in a 'revenue neutral' way. There is little point in putting so much effort in harmonization if other issues of political acceptability are taken into account. We hope our discussion reveals that more flexibility, leaving more room for subsidiarity, does not necessarily imply lack of decisiveness.

Notes

[1] Whether climate change implies global warming is still uncertain. As Schelling (1992) notes for instance, temperatures in some localities may even be reduced by the changes in climate, even if temperatures rise on average. For a concise survey of the scientific aspects see Cline (1991).

[2] Note this table uses CO_2 instead of C emissions; the conversion factor is 1 ton of C for 3.65 tons of CO_2.

[3] Various country-specific factors in the reference scenario contribute to these differences. For example, CO_2 emissions in the UK are expected to fall due to a shift from coal to natural gas, which contains far less carbon per unit of energy.

[4] Originally proposed in SEC(91)1744 (Comission, 1991), and formalised in a conditional proposal for a directive as COM(1992) 226 (Comission, 1992).

[5]The carbon/energy tax proposal was rejected by the heads of state and government of the Member States, at their summit in Essen (December, 1994). Instead, priority should be given to voluntary initiatives by individual Member States, including the introduction of carbon/energy taxes, within a common framework to be developed by the Commission and the Counsil of Ministers of Finance. On 28 February 1995 Energy Commissionar Papoutsis declared to the European Parliament that the Commission will not formally withdraw it's tax proposal. The fascilitation of Member State initiatives is now seen as a short-term, transitional solution, with a common EU carbon/energy tax as the prefered long-term goal. See also the working paper of the European Commission, *EU Climate Change Strategy: A Set of Options* (March, 1995).(Europe Environment, Nrs. 444/445, December, 1994; Global Environmental Change Report, Vol. 7, Nr. 4, Feb. 1995).

[6] See, for example, Pearson and Smith (1991), Smith and Vollebergh (1993), and Vollebergh (1994b) on which our analysis is based. The papers in the Special Edition of *European Economy* on 'The Economics of Limiting CO_2 Emissions', 1992, provide an excellent background analysis of the European carbon abatement case.

[7] Nevertheless, several other arguments can justify the energy component of the tax. First, the energy component may be intended to encourage energy efficiency, for reasons unrelated to reduction of carbon dioxide emissions. Second, the increase in energy prices is more similar across Member States leaving intra-EU competition much the same. For a further discussion of this issue, see Pearson and Smith (1992).

[8] In the next section we show that a carbon tax of 20 ECU per ton CO_2 will result in roughly 25% lower abatement costs for the EU as a whole. Nevertheless Denmark, Greece, Italy, Portugal and Spain would face higher abatement costs, compared with a uniform reduction of emissions by all Member States.

[9] For simplicity, carbon is chosen as both the tax base and the permit base in this and the next section. The tax or permit base could easily be replaced by a combination of energy and carbon.

[10] This so-called 'double dividend' issue spawned an enormous number of papers discussed in Goulder (1994). See also Chapter 8 of this volume.

[11] This explains why our calculations are not a simulation of the proposed EU tax.

[12] As noted by Bohm (1994), many evaluations comparing carbon tradeable permits with carbon taxes too often start from the alleged presumption that tradeable permits are allocated using grandfathering, and therefore do not raise revenue for the government. This asymmetry is due to current practice in the US where the existing experience uses this tradition all the time.

[13] The incentive for firms to relocate abroad is not removed completely, for a grandfathered tradeable permit has a certain value as a capital asset on the tradeable permit market. However, in this system the relocation tendency is self-defeating, because it pushes the permit price down.

[14] This section is based on Koutstaal, Vollebergh and De Vries (1995), who provide a more extensive analysis of the different systems involved.

[15] As far as we know this proposal was put forward in the Netherlands for the first time some years ago, when the different options for implementation of a Dutch energy/carbon tax were discussed

(see SER, 1993). In other policy areas the idea of tax thresholds is well-known (e.g. in labor taxation).

[16] A disadvantage of an akin tax-subsidy scheme proposed by Pezzey (1992) is that the authorities have to pay subsidies. According to this proposal polluters receive a subsidy for every abated emission under their allotted emission rights, while they have to pay tax for every ton emitted above this level. Consequently, the instrument might well cause higher government expenses, resulting in another problem for political acceptability.

[17] The energy tax in the Netherlands, to be introduced in 1996, only charges energy consumption exceeding 800 m^3 natural gas and 800 kWh electricity per household. Interestingly, only an estimated 5% of the households uses less gas or electricity.

[18] It should be noted that Member States are not allowed to use their permits to protect certain industries. Grandfathering is only allowed according to general EU rules, otherwise it would be in conflict with EU legislation on state support (art. 90, Maastricht Treaty).

References

Barrett, S. (1992a), 'Transfers and the Gains from Trading Carbon Emission Entitlements in a Global Warming Treaty'. In: UNCTAD, *Combatting Global Warming, Study on a Global System of Tradeable Carbon Emission Entitlements*, United Nations, New York.

Barrett, S. (1992b), 'Reaching a CO_2 Emission Limitation Agreement for the Community: Implications for Equity and Cost-effectiveness'. In: European Economy, *The Economics of Limiting CO_2 emissions*, Special Edition, Brussels, pp. 3-24.

Baumol, W.J. and W. Oates (1988), *The Theory of Environmental Policy*, Cambridge University Press, Cambridge.

Bohm, P. (1994), *Government Revenue Implications of Carbon Taxes and Tradeable Carbon Permits, Efficiency Aspects*, paper presented at the 50th IIPF Congress, Harvard, Cambridge MA, August 1994.

Bovenberg, A.L. (1993), 'Policy Instruments for Curbing CO_2 Emissions: The Case of the Netherlands', *Environmental and Resource Economics*, Vol. 3, pp. 233-244.

Broome, J. (1992), *Counting the Cost of Global Warming*, The White Horse Press, Cambridge.

Brouwer, O.W. (1994), 'Subsidarity as a General Legal Principle'. In: M. Dubrulle (ed.), *Future European Environmental Policy and Subsidiarity*, European Environmental Issues, Nr. 1, European University Press, Brussels.

Cline, W.R. (1991), 'Scientific Basis of the Greenhouse Effect', *The Economic Journal*, Vol. 101, pp. 904-919.

Cnossen, S. and H.R.J. Vollebergh (1992), 'Toward a Global Excise on Carbon', *National Tax Journal*, Vol. 45, Nr. 1, pp. 23-36.

COHERENCE (1991), *Cost-effectiveness Analysis of CO_2 Reduction Options. Report for the Commission of the European Communities*, DG XII, May.

Commission of the European Communities (1991), 'A Community Strategy to Limit Carbon Dioxide Emissions and to Improve Energy Efficiency', SEC (91) 1744, Brussels.

Commission of the European Communities (1992), 'Proposal for a Council Directive Introducing a Tax on Carbon Dioxide Emissions and Energy', COM (92) 226 final.

Dwyer, J.P. (1992), 'California's Tradeable Emissions Policy and its Application to the Control of Greenhouse Gases'. In: OECD, *Climate Change - Designing a Tradeable Permit System,* Paris.

European Economy (1992), *The Economics of Limiting CO_2 emissions,* Special Edition.

Goulder, L.H. (1994), *Environmental Taxes and the Double-Dividend: A Readers Guide,* paper presented at the 50th IIPF congress, Harvard, Cambridge MA, August 1994.

Hahn, R.W. (1989), 'Economic Prescriptions for Environmental Problems: How the Patient Followed the Doctor's Orders', *Journal of Economic Perspectives,* Vol. 3 (Spring), pp. 95-114.

Heister, J., P. Michaelis and E. Mohr (1992), 'The Use of Tradable Emission Permits for Limiting CO_2 Emissions'. In: European Economy (1992), *The Economics of Limiting CO_2 Emissions,* Special Edition, Brussels, pp. 27-62.

Hoel, M. (1991), 'Global Environmental Problems: The Effects of Unilateral Actions Taken by One Country', *Journal of Environmental Economics and Management,* Vol. 20, pp. 55-71.

Hoeller, P. and J. Coppel (1992), *Energy Taxation and Price Distortions in Fossil Fuel Markets: Some Implications for Climate-change Policy,* OECD Working Papers, Nr. 110, Paris.

Koutstaal, P.R. (1992), *Verhandelbare CO_2 Emissierechten in Nederland en de EG (Tradeable CO_2 Permits in the Netherlands and the EC),* Dutch Ministry of Economic Affairs, The Hague.

Koutstaal, P.R., H.R.J. Vollebergh and J. de Vries (1994), *Hybrid Carbon Incentive Mechanisms for the European Community,* Centre for Economic Policy (OCFEB), Research Memorandum 9406, Erasmus Universiteit Rotterdam.

Larsen, B. and A. Shah (1994), 'Global Tradeable Carbon Permits, Participation Incentives, and Transfers', *Oxford Economic Papers,* Vol. 46, pp. 841-856.

Nordhaus, W.D. (1991), 'To Slow or Not to Slow: The Economics of the Greenhouse Effect', *The Economic Journal,* Vol. 101, pp. 920-937.

Pearson, M. and S. Smith (1992), *The European Carbon Tax: an Assessment of the European Commission's Proposals,* The Institute for Fiscal Studies, London.

Pezzey, J. (1988), 'Market Mechanisms of Pollution Control: 'Polluter Pays', Economic and Practical Aspects'. In: R.K. Turner (ed.), *Sustainable Environmental Management: Principles and Practice,* Belhaven Press, London, pp. 190-242.

Pezzey, J. (1992), 'The Symmetry between Controlling Pollution by Price and Controlling it by Quantity', *Canadian Journal of Economics,* Vol. 25.

Poterba, J.M. (1991), 'Tax Policy to Combat Global Warming: On Designing a Carbon Tax'. In: Rudiger Dornbusch and James M. Poterba (eds.), *Global Warming,* The MIT Press, Cambridge Mass., pp. 71-98.

Rose, A. (1992), 'Equity Considerations of Tradeable Carbon Emission Entitlements'. In: UNCTAD, *Combatting Global Warming, Study on a Global System of Tradeable Carbon Emission Entitlements,* United Nations, New York.

Schelling, T. (1992), 'Some Economics of Global Warming', *American Economic Review,* Vol. 82 (March), pp. 1-14.

SER (1993), *Advies over de Invoering van Regulerende Energieheffingen* (Advice on the Implementation of Regulating Energy Charges), The Hague.

Smith, S. (1992), 'Taxation and the Environment: A Survey', *Fiscal Studies*, Vol. 13, Nr. 4, pp. 21-57.

Smith, S. and H.R.J. Vollebergh (1993), 'The European Carbon Excise Proposal: A Green Tax Takes Shape', *EC Tax Review*, Nr. 4, pp. 207-221.

Tietenberg, T.H. (1985), *Emissions Trading: An Exercise in Reforming Pollution Policy*, Resources for the Future, Washington, DC.

Tietenberg, T.H. (1992a), 'Relevant Experience with Tradeable Entitlements'. In: UNCTAD, *Combatting Global Warming, Study on a Global System of Tradeable Carbon Emission Entitlements*, United Nations, New York.

Tietenberg, T.H. (1992b), 'Implementation Issues: a General Survey'. In: UNCTAD, *Combatting Global Warming, Study on a Global System of Tradeable Carbon Emission Entitlements*, United Nations, New York.

Vollebergh, H.R.J. (1994a), *Environmental Taxes and Transaction Costs*, Tinbergen Institute Discussion Paper 94-96, Erasmus University, Rotterdam.

Vollebergh, H.R.J. (1994b), 'Transaction Costs and European Carbon Tax Design'. In: M. Faure, J. Vervaele, and A. Weale, *Environmental Standards in the European Union in an Interdisciplinary Framework*, MAKLU, Antwerpen, pp. 135-154.

WRR (1992), '*Milieubeleid, Strategie, Instrumenten en Haalbaarheid*' (Environmental Policy, Strategy, Instruments and Acceptability), SDU Publishers, The Hague.

Zodrow, G.R. (1992), 'Grandfather Rules and the Theory of Optimal Tax Reform', *Journal of Public Economics*, Vol. 49, pp. 163-190.

4 A European Solvent Tax to Reduce VOC Emissions

Xander Olsthoorn
Frans Oosterhuis
Frans van der Woerd

Introduction[1]

Volatile organic compounds (VOCs) is the collective term for a large variety of natural and man-made substances, with a wide range of applications. Accordingly, the number and diversity of sources of VOC air pollution is also large. Total anthropogenic VOC emissions in the EU amount to approximately 10 million tons per year (excluding methane) (Allemand *et al.*, 1990a). About a third of these (3 million tons) can be attributed to the use of (organic) solvents (Table 4.1).

Air pollution from solvent VOCs may have both a direct and an indirect effect on human health and the environment. Direct effects are due to the toxicity of VOCs found in solvents, whereas indirect effects are due to the fact that VOCs are precursors in ozone formation occurring both locally (episodic ozone) and over large areas (trans-boundary air pollution).

The Fifth Environmental Action Programme of the EU sets as a target for VOC policy a reduction of man-made VOC emissions by 10% in 1996 and 30% by the year 2000, based on the 1990 level. The EU wants to develop policy instruments to address air pollution from solvent VOCs. Air pollution from large sources will be covered by a directive, the *Solvent Directive*, providing emission limits for certain industrial processes. Such an approach appears inappropriate in the case of small sources or consumer products containing solvents. In those cases resorting to economic instruments seems a proper approach.

The use of organic solvents is subject to regulation for various reasons. The oldest regulations have been set because of the toxic properties of a number of specific solvents. Various organic compounds, which are used or have been used, are carcinogenic, teratogenic and have adverse effects on the human nervous system. Policies addressing these problems also belong to the realm of public and industrial health. This, together with environmental concern about the contribution of these emissions to formation of ambient ozone, constitutes, in brief, the background of the EU policy to control emissions of organic solvents.

Table 4.1: Solvent emissions (1000 tons) by end-users (EU level, situation 1985).

Sector	Emissions
Solvent Directive' firms	
Printing	191
Surface cleaning	493
Automobile manufacturing	147
Other coating (metal articles/wood)	558
Dry cleaning	84
Other firms	
Architectural paints	365
Other painting (marine, etc)	138
Manufacturing processes using solvent	453
Application of glues and adhesives	183
Households	
Architectural paints	103
Other (cosmetics, cleaning agents, other)	403
Total	3118

Ozone in the lower atmosphere is formed from organic compounds and NO_x through a complex of photochemical reactions. According to characteristic time scales of the process of ozone formation and of meteorological processes, three types of environmental ozone problems may be distinguished:

- urban ozone (diurnal variation): local scale, mainly related to VOCs and NO_x from traffic;
- episodic ozone formation: high ozone levels during weather conditions favorable for ozone build up (stagnant weather for several days) at a regional scale;
- background ozone formation: ozone in rural areas, trans-boundary issue.

A distinction between different types of ozone problems is relevant because different organic compounds add differently to these ozone problems: the various organics have diverging photochemical reactivities. However, these differences disappear when considering long-term contribution to background ozone formation. All VOCs in the air, eventually, contribute to background ozone formation. This is an important conclusion because reduction of VOC emissions is not necessarily an effective approach to control episodic ozone formation. This is due to the fact that ozone is produced from both VOCs and NO_x via mechanisms which imply that under certain conditions (large VOC/NO_x ratio) VOC emission reduction is not effective in mitigating ozone production on a time scale of days (episodic). These conditions prevail in countries where VOC emissions from natural sources (forests) are important. In those situations only NO_x emission reduction is effective. In the case of small VOC/NO_x ratios the situation is reversed: only VOC emission reduction is effective.

In this paper, the summarized results of a study are presented, which the authors have carried out for the EU, regarding the feasibility and implementation aspects of economic instruments to reduce VOC emissions from solvents which are used in small stationary sources and products. In the next section, the question will be addressed which type(s) of economic instrument(s) would be the most adequate for the issue at hand. The relationship between the economic instrument and direct regulative measures will also be dealt with.

The succeeding section deals with the particularities of the instrument which seems the most feasible, namely a solvent tax. The following issues are discussed: (1) the choice of the taxable object and the tax base, (2) the assessment of the tax rate which would be necessary to achieve a substantial VOC emission reduction, (3) the moment of levying, the taxable subjects, the exemptions to be made, the question of whether or not to earmark the revenues from a solvent tax, and the international trade aspects. The fourth section is devoted to the effectiveness of a solvent tax and its economic consequences, based on an assessment of the vulnerability of the different branches of industry which will be affected. In the last section, finally, some conclusions will be drawn.

The Choice of Instrument

In March 1993, the European Commission presented a draft for a Directive on the limitation of VOC emissions due to the use of organic solvents in certain processes and industrial installations (the 'Solvent Directive'; Commission of the European Communities, 1993). This Directive only relates to the relatively large stationary sources. Its main features are that it requires the application of BATNEEC techniques in about ten categories of production installations (printing, degreasing, coating, car manufacturing and other) and the introduction of solvent management plans in the firms concerned.

According to the Commission, setting specific emisson limits for small stationary sources and products would be less appropriate, mainly due to control problems and high costs. For these sources, the use of economic instruments is being proposed. This proposal is in line with the recent increase in interest for economic instruments in the environmental policy of the EU. The EU's wish to 'get the prices right' has been expressed in the Fifth Environmental Action Plan (Commission of the European Communities, 1992a, p. 67) and has resulted, among other things, in a draft Directive for a charge on the emission of carbon dioxide and the use of energy (Commission of the European Communities, 1992b).

When it comes to the choice of economic instrument, it seems clear that the two main candidates are taxes (fees, charges) and tradeable permits. Other economic instruments, like subsidies, fiscal incentives, deposit-refund systems[2] and liability may easily conflict with the Polluter Pays Principle (PPP), and/or are only feasible in certain specific cases or applications.

A recent OECD study (OECD, 1991) lists the relative advantages and appropriate circumstances for different types of economic instruments. From this 'checklist' it can be concluded that both a product charge and marketable permits have clear advantages and may raise some specific issues at the same time. Moreover, the relative advantage of either instrument may differ by the type of solvent and application. Theoretically, charges are to be preferred if the elasticity of demand is relatively high and the slope of the damage curve is not very steep (i.e., the risks involved in failing to attain the desired emission level are not disproportionally high). Marketable permits, on the other hand, should be chosen if exceeding the emission ceiling would result in high damage costs, while the elasticity of demand is low. However, damage costs and price elasticity are not the same for all solvents, and will also differ between various applications of one solvent. A separate system for each solvent and its application is, of course, out of the question. Therefore, the choice of the system may have to be made on more pragmatic grounds.

It is unlikely that marketable permits for solvent production for use in the EU can be introduced in the short term. While product charges are well established throughout the European Union, most Member States have no experience with tradeable production or sales rights. New legislation and institutions will have to be introduced which may take at least a few years. Furthermore, it is doubtful whether agreement could be reached on the initial distribution of such rights. Allocation based on historical production levels would punish the producers who had already adopted low-solvent production methods or products before the reference date, whereas an auction would create the risk of large producers trying to seize monopolistic or oligopolistic market power.

In the remainder of this paper, we will therefore analyze the different aspects of an instrument which raises the price of (products containing) solvents. We will call this instrument a solvent tax, because the term 'charge' suggests a direct relationship to emissions (which will not always be the case), and a 'fee' has the connotation of a payment for certain services rendered by the authorities.

Features of a Solvent Tax

The Taxable Object and the Tax Base

The goal of the economic instrument is to reduce the emissions of VOCs from small stationary sources and products. It is, however, quite obvious that a charge based on the actual emissions from these sources is not feasible, as the cost and effort involved in monitoring and enforcement would be insurmountable. In fact, the mere reason for introducing an economic instrument is to avoid the large effort and costs involved in monitoring all these small sources. A more practical solution would be to relate the tax to the amount of solvents produced, imported, sold or used. In that case, the objects to be taxed can be either the solvent as such, or products containing solvents. In order to maintain a closer relationship between the tax paid and the VOCs emitted, some kind of exemption or refund scheme could be introduced for those who can show that the solvents they use do not result in air emissions (see below).

Alternatively, one could try to make a distinction between small stationary sources on the one hand, and final products on the other hand. An emission charge could then be levied from the stationary sources, using some kind of proxy parameters to calculate the emission (e.g., the amount of solvents used and the presence of recycling and/or emission treatment equipment), while the final products would be charged with a product tax. However, this would lead to higher administrative costs and possibly to problems in cases where the two systems overlap. It therefore seems preferable to use the same system for products and stationary sources. Such a system could either take all imported and produced solvents as the taxable object, or certain final products containing solvents. The latter alternative has the advantage that it can concentrate on those products which are mainly used by consumers and firms which do not come under the Solvent Directive. On the other hand, it could lead to unwanted substitution by non-taxed products.

A tax on solvents can be levied *ad valorem* or per kg of substance produced, imported, etc. An *ad valorem* tax is a constant proportion of the value of the product, which makes it insusceptible to inflation. However, it places an artificially heavy burden on the most expensive kinds of solvents, which would only incidentally be justified from an environmental point of view. As the present inflation rates in the EU are not so high as to make very frequent adjustments of the charge rate necessary, the amount of solvents (expressed in kilograms) seems to be the most appropriate tax basis.

The Tax Rate

A product tax could be uniform for all solvents, or differentiated according to the properties of the substance. A uniform tax is administratively simple, but does not take the specific environmental properties of each substance into account. Given the existing classification system, a differentiation would be feasible in two broad classes, e.g. a class of carcinogenic and chlorinated solvents (with the highest rate) and a class of other solvents. Such a differentiation is present in the Swiss proposals for a VOC tax (Bruhin *et al.*, 1988; 1989). A similar distinction has also been made in the proposed EU directive on solvent emissions (Commission of the European Communities, 1993).

Another possible differentiation would be a geographical and temporal one. As the environmental damage caused by VOC emissions is very much dependent on the exact time and location of the emission, such a differentiation would be desirable from a theoretical point of view. In practice, however, this is not feasible, because one cannot predict where and when a certain amount of solvent, which is subject to the tax, will lead to VOC emissions. Even a differentiation between Member States would probably turn out to be unworkable, because it could easily be frustrated by cross-border shipments. Although in principle a differentiation would be possible as long as it is lower than the associated transport costs, we will not consider a regionally differentiated tax rate.

For the sake of completeness, the possibility of introducing a threshold and a progressive rate in the tax should be mentioned. This would spare the small users and burden the large ones relatively heavily. However, it seems quite clear that such a system would be administratively unfeasible and susceptible to fraud.

The Solvent Directive requires firms to apply BATNEEC techniques in new plants and in existing plants after a specified period. In line with this phased introduction a feasible method of softening the (economic) blow of a solvent tax for industry would be a gradual increase of the tax rate, starting at a relatively low level. This would enable solvent users to adapt their processes, products and preferences at relatively low costs. The final tax rate, to be reached in 10 to 15 years' time, should be announced well in advance, and the scheduled increases should be strictly adhered to.

The needed level of the tax rate will be determined by the emission reduction target and the price elasticity of the demand for solvents. Previous studies (Bruhin *et al.*, 1989; LMO, 1990; USEPA, 1991; Swedish Environmental Charge Commission, 1990) suggest a tax rate somewhere between 0.5 and 3 ECUs per kg for 'ordinary' VOCs, and between 1 and 6 ECUs per kg for chlorinated (and carcinogenic) VOCs. The associated emission reductions are of the order of magnitude of 30 to 60%.

The Moment of Levying: Taxable Subjects and Exemptions

The environmental effectiveness of the solvent tax will be enhanced if the tax is related as closely as possible to the amount of VOCs actually emitted. From this point of view, it would be preferable to levy the tax at the 'end of the chain', i.e. from the final user of the (product containing) solvent. However, the number of solvent users is far too great to make this a feasible option.

A more workable alternative seems to be to impose the tax when the solvent (or the product containing solvents) is brought onto the market in the EU for the first time, i.e. when it is being sold by the producer or by the importer. This is the system used for many of the existing excise taxes. It has one clear and predominant advantage: the number of producers and importers, liable to pay the tax, is fairly small, thus restricting the administrative and enforcement costs. In this system, the producing or importing firm has the status of a 'bonded warehouse': it has a license to produce, store, receive and ship the taxable goods without being liable to pay the tax. The tax is due at the moment when the goods are being transferred to a second market party. A disadvantage of this system would be its lack of compliance with the 'destination principle' and the fact that the tax burden is also put on solvents which will not lead to VOC emissions (for instance, because they are being used as a feedstock or as a fuel). However, exemptions for this can be made (see below), the costs and administrative effort of which are probably much lower than those of a more complex tax system. We will therefore investigate only this excise type of solvent tax.

In order to be an effective, efficient and equitable instrument, the solvent tax rules should allow for some exemptions. First of all, *exports* of solvents and products containing solvents to non-EU countries should be exempted. If the tax has already been paid, it should be refunded at the border. This exemption can be justified from the point of view that domestic producers should not be put in a disadvantageous competitive position with regard to foreign producers who do not face the same kind of tax. Furthermore, the tax is meant to be an instrument for VOC emission reduction in the EU and not to improve the environment elsewhere in the world.

Secondly, the tax should not be levied if the substance is *not being used as a solvent*, but as a feedstock (e.g., for plastics) or as a fuel. The tax regulations have to contain provisions which allow for this distinction. It can be left to the Member States to decide if the user of the substance should be charged with the onus of proof that its application is of a non-solvent character, or if the administrator has to prove the opposite. This is largely a matter of what is regarded as decent administration. A general exemption could be made for substances like propane and butane, for which its use as a solvent is the exception rather than the rule.

Thirdly, *recycled solvents* should be exempted in order to comply with the Polluter Pays Principle and to stimulate recycling. Firms which only deal with recycled solvents should therefore not be obliged to register. Firms selling both virgin and recycled solvents should be able to prove by their administration which part is recycled and therefore exempted from the tax.

One might also consider making an exemption for those (applications of) solvents for which *no reasonable alternatives* are presently available. This would mitigate the financial burden for those users who have no alternatives at their disposal. However, it would also take away the incentive to search for alternatives, which is one of the main functions of the solvent tax. Furthermore, if such an exemption would mean that the solvent use concerned remains unregulated, this would conflict with the Polluter Pays Principle. Finally, a lower rate for certain applications only will be administratively complex and susceptible to fraud.

The number of solvent applications where equivalent alternatives are lacking is rather limited. The impact of a solvent tax for these applications depends on the elasticity of demand for the product or service for which the solvent is being used. If this elasticity is high (in absolute terms), demand will fall sharply with the price increase caused by the tax. In that case, the producer will bear the cost of the tax (in terms of market loss). This situation will probably occur only in the case of a few 'luxury' products, like some automotive products and cosmetics (although the demand for e.g. expensive perfumes may also be quite price-inelastic). If the price elasticity of demand is low, the consumer will bear the cost of the tax (in terms of higher prices). This situation may occur in some special applications, where the result (of degreasing, for instance) has to meet high standards.

One of the big advantages of a tax over a ban is the fact that the market will determine where solvents are really indispensable. This precludes lengthy disputes between administrators and firms over the question of which applications should be exempted and which ones should not. Should one nevertheless, for political reasons, want to make such exemptions, then it will be necessary to define exactly the applications to be exempted. The user will have to prove that he is using the solvent or the product for one of these applications. This would probably involve considerable administrative efforts and, again, create possibilities for fraud.

A tax refund can be considered for cases where the user can demonstrate that effective *measures have been taken to prevent the solvent from leading to air emissions.* Several options for emission reduction are available: the solvent can be recycled, collected for reprocessing, captured in some 'end-of-pipe' device (e.g., a biofilter) and subsequently treated as chemical waste, or it may be burnt in an incinerator. The decision whether or not to exempt these activities and treatments from the tax is mainly a trade-off between the expected incentive effect on emission reduction and the administrative costs and opportunities for fraud.

In the case of recycling or reprocessing, a tax refund should not be given, because the proposed system already envisages a tax exemption for recycled solvents. As long as the solvent has not been destroyed, it has the potential to produce VOC emissions and should therefore bear the tax.

When solvents are incinerated or otherwise treated as waste, VOC emissions are replaced by other emissions (CO_2, chemical waste), which could justify an environmental tax as well. If exemptions for these kinds of treatment are nevertheless made, it is important to state exactly the requirements for proving that the emission has been prevented. It is, for instance, conceivable that the presence of emission control equipment is not proof enough, but that one should also be able to show that it is actually and effectively being used. Obviously, the solvent management plans - required according to the Solvent Directive - could be used for this end in the case of those sources which are the subject of the Directive.

Finally, in our view, there should be no exemption from the tax for *industries which are subject to the EU Solvent Directive.* Although the economic instrument is primarily meant to reduce the VOC emissions from small stationary sources and products, it seems quite obvious that a tax on solvents can not be restricted to these categories alone. The tax will also be charged on solvents used in large stationary sources. This means that these large sources will be subject both to direct regulation (according to the Solvent Directive) and to the tax. In other words, the sources which come under the Directive carry a heavier regulatory burden: they do not have the choice between paying and polluting (like the small sources and product users), but they are not compensated for that financially. If this is felt to be undesirable, one could select those products for taxation which are mainly used by 'non-Directive' firms and households. If the tax were to be levied on all solvents, special exemption rules would have to be introduced for the Directive part of industry. This may, however, lead to distortions and create opportunities for tax evasion and fraud. For example, several small installations could be combined to make up a large one, or an untaxed batch of solvents could easily be used in 'taxable' applications. Furthermore, it seems only fair to tax the remaining solvent use (emissions) of large sources if smaller sources have to pay for it also. Finally, the tax will in practice be an incentive for many enterprises to reduce their solvent emissions even further than the Solvent Directive prescribes. In such cases, the limit values of the Directive cease to be a binding restriction for the firm; this role is being taken over by the tax. We will therefore assume in our analysis that special regulations for sources which fall under the Directive will not be applied.

Destination of the Tax Revenue

In discussions about environmental taxes and charges, the question of earmarking the revenues is always a hot item. Whereas the presently existing environmental levies have generally been introduced to finance specific purposes, the proposed tax on solvents is meant to be a pure incentive tax and its revenues can be seen as a side-effect. Therefore, there is no reason *a priori* to allocate this money to environmental funds. Moreover, as several Member States have put a ceiling on their total tax burden, the solvent tax will have to be compensated by a decrease in some other tax (fiscal neutrality). If the proceeds of the solvent tax were reserved for the environmental budget, this would mean a forced reallocation of public funds. Finally, the solvent tax will not, by its very nature, be a stable source of government income. In the case of earmarking, a drop in tax revenues due to a successful solvent tax would hit the environmental budget unilaterally.[3]

A major reason why earmarking for environmental purposes should nevertheless be considered, is the fact that it might lead to greater political and public support for the tax. People tend to be less opposed to taxes if the revenue is used for a purpose to which they are sympathetic, and which bears some relationship to the taxed object. One example might be the establishing of a (temporary) fund to finance R&D projects in the area of solvent reduction, stimulate the diffusion of best available techniques (BAT), compensation payments for damage caused by photochemical smog, etc. In the case of a tax on final products, the revenues could also be used to subsidize their low-solvent alternatives (this is possible because the producers have to register their sales of the taxable products and their solvent contents). This constitutes a second reason for earmarking: the rate of reduction of VOC emissions can be increased and (cumulative) costs for firms lowered, as low-VOC techniques or products will enter the market earlier.

It has also been suggested that the tax revenues be refunded (after deducting administrative costs) to those who bear the burden of the tax. However, in the case of a solvent tax it is unclear who should be the beneficiaries of such a redistribution, and what the distributive code should be.

Finally, one might consider the possibility of redistributing the revenues among the Member States in order to allow for differences in the effects of VOC emissions (due to differences in other (anthropogenic and natural) emissions, climatic conditions and exposed objects). For example, a country like Spain would get only minor environmental benefits from a reduction in VOC emissions. If, for practical reasons, the solvent tax were levied in Spain at the same rate as elsewhere in the EU, a financial compensation could be given from the tax revenues. Maybe the Cohesion Fund could play a role in this matter.

The Solvent Tax and International Trade

In the proposed tax system, all solvents and products containing solvents are treated equally, irrespective of their origin (imported or domestically produced, that is, within the EU).[4] Therefore, the tax will not conflict with the relevant GATT/WTO rules. It would be a different story if the tax were also applied to products which have been *treated* with solvents. This would interfere with the autonomy of the exporting country to pursue its own environmental policy, as the emission takes place at the production site.[5] Some borderline cases might be found where products contain solvents which are being gradually released (such as certain synthetic foams or preserved wood). The emissions from these products take place partly in the country of origin and partly in the country of destination. To prevent problems, it would be wise not to include such products in the list of taxable goods.

The solvent tax could create an incentive for some firms to transfer their solvent-based activities (e.g., metal degreasing) to countries where such a tax is not applied. The scope for this seems rather limited, however, as this will only be profitable when the difference in solvent costs compensates for the higher transport costs.

The taxation of imported products implies that their solvent content should be indicated on the product itself, or in an accompanying document. Although this requirement could be regarded as a technical barrier to trade, it seems clear that it has no protectionist intention at all, and therefore does not violate the GATT/WTO Standards Code. A similar obligation for products manufactured within the EU would not be necessary for the tax as such, but could nevertheless be considered in order to improve the feasibility of solvent management plans. As the ECE protocol on VOC emissions also provides for product labelling, coordination efforts should preferably take place at that level.

Environmental and Economic Effects of a Solvent Tax

The Actors in the Solvent Chain

An overview of the flows of solvents - from production to final use - and the main economic actors involved is presented in Figure 4.1.

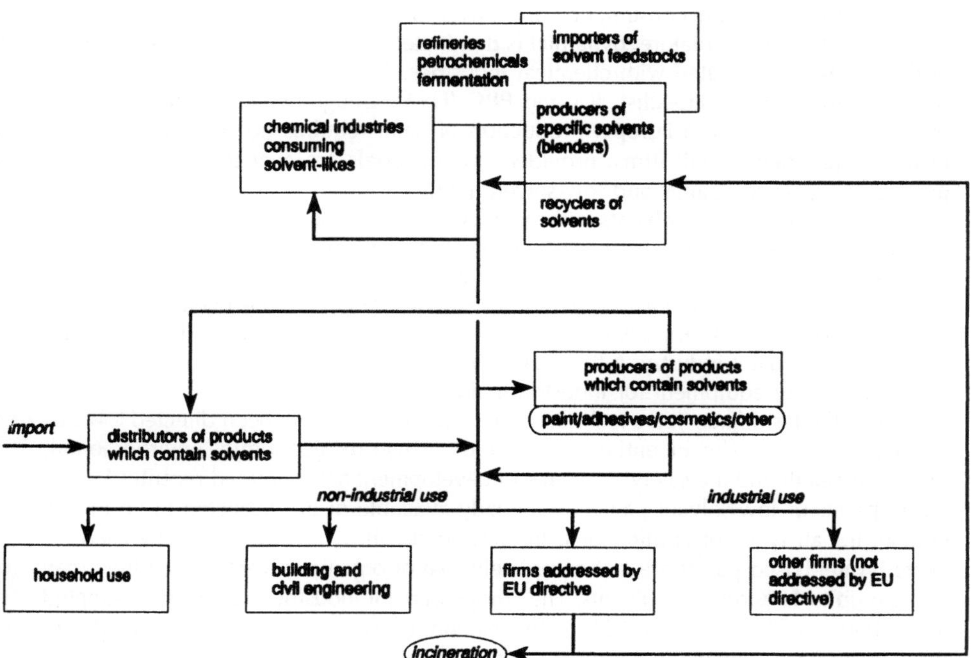

Figure 4.1: Flows of solvents and economic actors.

The production of organic solvents - The firms involved in the production and marketing of solvents may be divided into four groups:

- the oil refineries, which produce mainly aliphatics and aromatics;
- the (petro)chemical industry, which produces:
  - aliphatics and aromatics as a by-product of steam-cracking;

· oxygenated hydrocarbons;
· halogenated hydrocarbons (chlorine-based industry);
- the industry which distills ethyl alcohol from fermented material (NACE 424);
- the recycling industry (mainly chlorinated hydrocarbons).

About half of the solvents are produced by the refining industry, mainly aliphatics and aromatics. The chemical industry produces a similar amount (oxygenated and chlorinated hydrocarbons). Less than 5% (ethyl alcohol) is produced from agricultural feedstock.

For oil refineries a reduction of the consumption of organic solvents will affect less than 0.5% of their physical output and turnover. The share of solvents in the turnover of the chemical industry as a whole is estimated at 1-2%. Within the chemical industry the impact of decreasing production will be unevenly distributed among the various firms, according to their involvement in solvent production. Some firms may gain from a decline in the use of VOCs, if they produce substitutes. Ethyl alcohol used as a solvent may constitute about 10% of the output (in volume) of the sector 'Distilling of ethyl alcohol from fermented material; spirit distilling and compounding'.

An increase in the market-value of solvents will stimulate both trade in spent solvents and the recycling industry.

The production of products which contain solvents - About half of the total amount of solvents (1.5 million tons) is used in coating, painting or similar applications. Therefore the *paint industry* is a key player in the VOC emission arena.

Both production and consumption of paints is concentrated in the Northern European countries. The market (mature, dynamic) is dominated by less than 10 large firms, some of them closely associated with chemical industries. Among them are major industries such as AKZO, BASF, Hoechst, ICI and PPG (USA). Geographically, small firms are predominant in Southern Europe. Production of paint requires no specialized know-how or equipment. Small firms, provided raw materials are available, can, and do, produce water-borne paints and low-solvent paints.

The share of organic solvents in the production costs of paints ranges from 0 to 5%. Related to value-added the share of solvent costs ranges up to 15%.

Paint users, both households and industries, are technologically dependent on this industry. However, the paint industry in turn is dependent on the chemical industry which markets the raw materials. On the other hand major chemical industries are involved in the paint industry. A third party, involved in industrial applications, are the manufacturers of equipment for industrial painting and coating.

Over the last fifteen years technological developments have been directed towards less use of VOCs. The incentives are various in nature: economic, environmental, industrial health and fire safety. The pace of development has increased recently. Water-based paints or low-solvent paints are already available or are expected to be in due course, for all types of applications. In some cases they have been widely used for years. Water-based paints are not completely free of organic solvents, since they may contain up to 12% organic solvents. The European paint industries (CEPE) have adopted a maximum VOC content of 250 g/liter (excluding water; about 10% by weight) in paint as an environmental mark for decorative paints.

For nearly all decorative paint (used in building and by households) low-solvent paints are available. The barrier for wide-scale use is market acceptance, constituted by different properties of low-solvent paints, risk-avoiding behavior of households and painting firms, lack of awareness and price differences.

The situation with respect to the *ink industry* (amount of solvents involved: 150 - 200 kilotons) is quite similar to that of the paint industry. One difference is that inks contain more expensive solvents (alcohols, esters, ethers and the like). Another difference is that the consumer market is hardly relevant to the ink industry.

The *cosmetics, toiletries and perfumes industry* ranks third on the list of industries putting solvents in their products. The total amount of VOCs annually processed is estimated at 100,000 tons. Ethyl alcohol is the main chemical used (in volume). It should be noted that not all the solvent is emitted into the air; part is washed away and enters into (waste) water. The share of solvents in the production costs is about 2%, while related to value added this share is 6%. These are averages over all products. Hairsprays are the main products with respect to emissions. In contrast to paint, lacquers and ink, the scope for reduction of the VOC content is limited.

The industry which produces *cleaning agents and polishes* will also be affected. This segment of the chemical industry manufactures products used in households, offices and in car maintenance. The range of products varies widely and may contain only small amounts of VOCs.

The final users of products with organic solvents - 'Solvent Directive' industries are those types of industries which might be subject to the proposed directive if their emission exceeds the level indicated in the directive. Major activities subject to the proposed regulation are painting and coating of various substrates, printing operations and metal degreasing. In several cases non-VOC technologies are available in metal degreasing and metal and wood product coating. Besides economic factors, the fact that production processes are often prescribed by the clients of (subcontracting) firms may also slow down technological change.

The *building and construction industry* is involved in the solvent issue primarily as a user of paint. Around 1985, the use of architectural paints accounted for about 365,000 tons of emission. Additional emissions are due to the use of glues, adhesives and cements.

The building industry is large in both turnover and number of firms. Painting is usually carried out by subcontractors: painting firms. The costs of solvents in the paint used constitute less than 1% of the value added of these firms.

Water-borne and low-VOC paints are available for a wide range of applications in building and construction. Compared with conventional paints they perform equally well, on the average. It is expected that performance and range of application will increase. In Denmark (and Sweden) the use of organic-solvent borne paint is restricted for occupational-health reasons.

Barriers to entering the market are the risk-avoiding behavior of building firms and the higher costs of low-solvent alternatives for organic-solvent paints.

Other users of products with organic solvents include various types of firms such as *garages and similar mechanical shops, offices, hairdressers and beauty shops.* Cleaning agents (degreasing solvents) and hair sprays are products wich contain or may contain large amounts of VOCs.

Households constitute the last sector dealt with. Architectural paints, thinner, cleaning agents and cosmetics (hairsprays) stand out on the list of household products containing organic solvents. The per capita costs of the solvents involved are estimated at a few (2-3) ECUs annually. The scope for emission decrease is mainly in the use of water-borne (or low-solvent) paints, also prompting a decrease in the use of thinner.

Environmental Effects

Table 4.2 summarizes our estimations[6] of the eventual emissions after implementation of a generic solvent tax of about 1 ECU per kg. The first column shows the current emissions. The second column shows the *reference scenario:* the emissions resulting from continuation of the current expected policies. Basic assumptions in calculating this scenario are:

- the Solvent Directive will be effective in a few years, and completely implemented by 2010;
- current initiatives in the area of communicative instruments (European eco-label, covenants between authorities and industry) will be continued;
- emissions (if the environmental policy mentioned above is not pursued) tend to be constant in the period 1990-2010 as economic growth rates offset the effects of autonomous trends towards low-solvent technology.

Table 4.2: Solvent emissions (1000 tons) by final users (EU level).

Sector	Situation 1990	Reference scenario 2010	Tax scenario 2010
'Solvent Directive' firms			
Printing	191	94	70
Surface cleaning	493	177	133
Automobile manufacturing	147	63	48
Other coating	558	324	243
Dry cleaning	84	21	16
Building & engineering			
Architectural paints	365	183	91
Other painting (marine, etc.)	138	104	52
Other firms			
Manufacturing processes using solvents	453	453	453
Application of glues and adhesives	183	137	69
Households			
Architectural paints	103	52	26
Other	403	302	151
Total	3118	1910	1350

The Fifth Environmental Action Programme of the EU has as a goal for the year 2000, to reduce the total VOC emissions (from all types of sources) by 30% in relation to 1990. It appears from the reference scenario that this reduction will not be reached before 2010. To attain the target by 2000 without additional measures, other sectors will have to contribute more.

The effect of a generic tax (*tax scenario*) on all solvents will differ from sector to sector and from solvent application to application. The impact can only be predicted with a reasonable amount of certainty for those solvents and applications for which equivalent substitutes, alternatives and/or emission reduction techniques are presently known, while the price or cost difference is the main obstacle to adopting them. In reality, these situations are rather rare. Several substitutes are already available (e.g. in the important area of painting), but higher prices are usually not the only reason for the slow adoption. The main roles of the solvent tax will be to speed up technological development towards low-solvent products and processes, and to influence the choice of producers and consumers (in which non-monetary considerations like performance, quality, safety and ease of operation play an important role as well). Trying to predict the outcome of these processes is a tricky undertaking. Our 'forecasts' should therefore simply be seen as tentative estimates.

For *firms subject to the Solvent Directive,* a tax will constitute an incentive to adapt technologies and procedures sooner than without a tax. Secondly, the incentive will favor either recycling or incineration depending on their respective costs and benefits (market value of used solvents in the case of recycling; refund of the tax in the case of incineration). The eventual emissions from these firms, subject to both the Solvent Directive and higher priced solvents will probably be only marginally less than without a tax. The function of the tax is mainly to support and to speed up the rate of application of environmental technologies.

Small 'Solvent Directive' types of firms (not subject to the Directive) are encouraged to take appropriate measures, such as good housekeeping and change to new technologies or perhaps abatement measures, although they can usually not afford end-of-pipe solutions because of economies-of-scale. However, the large number of small firms coping with increased costs of solvents constitutes a market for producers of low-solvent technologies. It is assumed that the additional reduction of emissions for both the large and small firms is 25%.

The main targets of a tax - in addition to the small 'Solvent Directive' types of firms - are *specific products which contain solvents* (paints, glues and adhesives, cosmetics, toiletries), firms which use small amounts of solvents, and the category remainder, consisting of countless products which contain VOCs. It is estimated that an additional 50% emission reduction will be achieved.

Economic Effects

Total emissions are not equivalent to the amounts of solvent marketed by the *solvent-producing industries.* Production is higher for the amount of solvents incinerated in pollution abatement equipment.[7] Incineration is an appropriate technique in the printing and coating sectors (including car manufacturing). It is estimated that production of (virgin) solvent in 2010 will be some 1600 ktons (about 50% reduction), while recycling will produce about 300 ktons. Solvents are usually produced by large chemical firms (such as Shell, ICI, BASF) and oil refineries. Solvents contribute to their turnover only to a limited extent. Low-solvent technologies may require other types of chemicals. Firms also engaged in the production of these substances may gain in the end.

Trade in (spent) solvents will suffer on the one hand as it appears that fewer solvents will be on the market. However, as the value increases, the turnover of these firms should not alter correspondingly, and may even increase. Recycling activities will obviously be promoted by an increase in the price of solvents.

Among the industries producing products which contain solvents, the *paint and ink* industry plays a leading part, as paint and lacquers are a major VOC emission source. Products which either do not contain organic solvents at all or only in small amounts are currently being developed. Differences in prices of solvent-based paints and their alternatives are not so high that demand for paints will decrease. An effect may be that large paint firms (owned by chemical firms) will increase their market share at the expense of small firms due to their technological lead. However, paint making itself is not a high-technology production process.

An increase in prices of *cosmetics and toiletries* is expected to be limited, implying also limited effects for the industry.

There will probably be major effects for (small) *firms which sell cleaning agents* based on VOCs (such as thinner and benzine).

On average, the effect of a tax on *final users of solvents* will be minor as the share of solvent costs in added value is usually below 1%. However, some small firms (e.g. specialized dry cleaners and manufacturers of metal products) could face serious problems. As these firms use mainly chlorinated solvents, this could be a reason to abandon the idea of a separate (higher) tax on this category.

Effects on the Public Budget

Fiscal authorities are usually not very fond of introducing environmentally oriented incentive taxes. First of all, such taxes are, almost by definition, not a stable source of income for the treasury. Secondly, the administrative costs of an ecotax are generally high, particularly if many exemptions and refund possibilities are introduced with a view to their equity and effectiveness as an environmental policy instrument. Finally, earmarking the revenues restricts the freedom of the policy makers to determine the destination of public funds, especially if, for reasons of budget neutrality, the introduction of the tax has to be compensated by reductions of other taxes.

Assuming a (uniform) tax rate of 1 ECU per kg on all solvents, introduced at once, the initial (potential) EU-wide revenue would amount to some 3 billion ECUs per year. Industry's response to the tax will lead to a decrease in solvent consumption, and thus revenues of (presumably) 60% by 2010 (see Table 4.2). The solvent tax revenues are of the order of 0.1% of total tax revenues in the EU countries.

Member States will have to appoint institutions responsible for the collection, control, enforcement and refund of the tax. The costs of performing these tasks depend in large part on the number of taxpayers. In case of a charge on solvents as such, the number of producers involved is about 100. Firms which trade products (paints, cosmetics) across EU borders are more numerous. We estimate this number to be no more than 1000. Furthermore, this system has to deal with applications for refunds from firms with end-of-pipe equipment. The number of firms eligible for such refunds can be estimated at no more than 1000 in the whole EU (although this figure could increase somewhat as a result of the tax). These are mainly large firms (primarily in the printing, metal coating and chemical industry) which will have to prepare 'solvent management plans' under the Solvent Directive. These 'solvent management plans' can serve as a basis for the refund. Assuming that checks will be primarily administrative and only occasionally physical, the total administrative costs under this system will probably not exceed 10 million ECUs per year, or less than 1% of the revenue.

If the tax is levied on certain product categories only, the number of tax payers is larger. In the EU there are, for instance, more than 2000 firms (with more than 20 employees) producing paint. The administrative costs will of course depend largely on the number of taxable products, but it is likely that they will be between 10 and 100 million ECUs. This is more than 1% of the expected revenue (also because of the narrower tax base).

If the principle of stability of the burden of public expenditure is adhered to, the introduction of the solvent tax will have to be accompanied by a reduction in other taxes. It can be left to the Member States to choose the tax they want to reduce. If the revenues of the solvent tax are earmarked, the result will be a lower revenue for the general budget. In that case, budget cuts (or deficit increases) are unavoidable.

Conclusions

Air pollution from VOCs due to the use of organic solvents poses a complicated policy problem. The number of solvents and applications is large. Environmental effects differ by solvent, but also by the time and location of the emission (meteorological conditions, presence of other pollutants). Some of the effects have a trans-boundary character, others are mainly regional or local. An economic instrument which takes all these considerations into account would imply prohibitively high costs of administration and enforcement. On the other hand, a 'simple' economic instrument (a uniform tax on all solvents without exemptions or refunds) will lead to inefficient allocation and inequities.

Clearly, some kind of compromise has to be found. In our view, the choice is between a tax on all solvents, or a tax on certain products containing solvents. The first alternative has the advantage of comprehensiveness and lower administrative costs. The second one avoids the need for refunds in the case of non-solvent applications and end-of-pipe treatment, and makes it possible to restrict the tax (mainly) to those sectors which are not already subject to direct regulation.

The tax rate should be approximately 1 ECU per kg solvent, possibly with a gradual increase over time and a higher rate for certain high-risk substances. The proposed tax is not likely to come into conflict with GATT/WTO trade rules. The economic impact on most industries involved will be modest.

The effects on VOC emissions of such a tax are not easy to forecast, as cost-effectiveness curves are generally lacking. Moreover, one of the functions of the tax is to speed up technological development towards low- and non-solvent products and processes. The outcome of these processes is of course unknown in advance. In other words: the cost-effectiveness curve will change as a result of the tax. Therefore, it is likely that a solvent tax will not only contribute directly to the EU targets of VOC emission reduction by stimulating the application of already existing technology, but also indirectly by stimulating innovations.

Notes

[1] An initial version of this paper was presented at the IIASA Conference on *Economic Instruments for Air Pollution Control*, Laxenburg, 18-20 October 1993. The research on which this paper is based was funded by the European Commisssion (DG XI).

[2] Macauley *et al.* (1992) suggest a deposit-refund system for chlorinated solvents, but they are mainly concerned with the final disposal of these solvents as chemical waste. In cases where dissipative applications are the main issue, they also recommend a product tax.

[3] This is not necessarily a problem if one assumes that the revenues are used specifically for VOC-reducing activities.

[4] At first, it may seem as if imported products containing solvents are being discriminated against, because there is a tax on imported products but not on domestic ones (unless they are brought on the market directly by the solvent producer). However, the tax on domestic final products has already

been paid when the solvent which the product contains was brought on the market for the first time.

[5] The famous 'tuna-dolphin case' between Mexico and the USA in 1991 provides the basic criteria which the GATT applies in matters like this.

[6] The available information about solvent emissions is notoriously uncertain (up to 100%). The assumptions about effectiveness of the different initiatives are crude. Therefore, the scenario may only, with care, be used for a trend analysis.

[7] Solvents may also enter a product and be released years later.

References

Allemand, N., *et al.* (1990a), *A Costed Evaluation of Options for the Reduction of Photochemical Oxidant Precursors. Volume 1: Results of Three Possible Scenarios for the Abatement of Photochemical Oxidant Precursors,* Commission of the European Communities, EUR 12537/I EN, Brussels, 1990.

Allemand, N., *et al.* (1990b), *A Costed Evaluation of Options for the Reduction of Photochemical Oxidant Precursors. Volume 2: Abatement Technology and Associated Costs,* Commission of the European Communities, EUR 12537/II EN, Brussels, 1990.

Bruhin, A., H. Meckel, J. Kaltenbach, P. Wirth, S. Wurm (1988), 'Entscheidungselemente für eine Lenkungsabgabe auf flüchtige Kohlenwasserstoffe', *Expertenbericht* (1. Teil), Basel.

Bruhin, A., S. Wurm, J. Kaltenbach, H. Meckel (1989), 'Entscheidungselemente für eine Lenkungsabgabe auf flüchtige Kohlenwasserstoffe', *Expertenbericht* (2. Teil), Basel.

Commission of the European Communities (1992a), *Towards Sustainability. A European Community Programme of Policy and Action in Relation to the Environment and Sustainable Development,* Volume II, COM(92) 23 final - Vol. II, Brussels, March 1992.

Commission of the European Communities (1992b), *Proposal for a Council Directive on the Introduction of a Charge on the Emission of Carbon Dioxide and on the Use of Energy,* COM(92)226 final, Brussels, June 1992.

Commission of the European Communities (1993), *Proposal for a Council Directive (EEC) on the Limitation of the Emissions of Organic Compounds Due to the Use of Organic Solvents in Certain Processes and Industrial Installations,* XI/240/93-FINAL, Brussels, March 1993.

Landelijk Milieu Overleg (LMO) (1990), *Financiële instrumenten voor het Nederlandse milieubeleid,* LMO, Utrecht.

Macauley, M.K., M.D. Bowes, and K.L. Palmer (1992), *Using Economic Incentives to Regulate Toxic Substances,* Resources for the Future, Washington D.C.

OECD (1991), *Environmental Policy: How to Apply Economic Instruments,* Organisation for Economic Co-operation and Development, Paris.

Swedish Environmental Charge Commission (1990), *Summary of Proposals,* SOU 1990:59.

United States Environmental Protection Agency (USEPA) (1991), *Economic Incentives - Options for Environmental Protection,* Office of Policy, Planning and Evaluation (PM-220), Washington, March 1991.

5 Economic Instruments in EU Waste Policy

Joram Krozer
Pascale van Duyse
Jan L. de Vries

Introduction

Waste recycling has been a priority in solid waste policy for the last few decades, though only a slight increase in the amount of recycling has actually been achieved. According to OECD data the recycling rate of household and industrial waste in the EU is less than ten percent of the total waste supply.

To encourage recycling, the application of product-oriented instruments is often proposed, such as deposit refunds, a premium for waste collectors, a fee on sales, and producers' responsibility for collection and recycling (e.g. Bohm and Russel, 1985; Russel, 1988; Pearce and Turner, 1992). However, with some limited exceptions, the implementation is difficult. Such instruments run up against difficulties like negative effects on sales, high additional costs of kerbside collection, administrative inefficiencies and complex implementation. So far they have only been occasionally implemented on a large scale by EU Member States (e.g. producers' responsibility in Germany).

The application of charges on waste disposal to encourage recycling has also been proposed (e.g. Brown, 1984). Some EU countries, like Denmark and Belgium, have introduced national waste charges with fairly low rates. The objectives of these charges are to sponsor recycling projects and/or to discourage landfilling of unprocessed waste. A stimulating effect on recycling is either not intended or rather insignificant because of the low level of the rates. The argument against much higher regulatory charges, introduced at the national level, is that at the same time they also provide incentives for increasing waste mobility. This can result in even more pollution because of the transportation of waste and uncontrolled disposal to cheap facilities.

So far, no definite proposal has been formulated to encourage effective waste recycling while avoiding the risks of large-scale waste mobility. In 1989, the EU made an important contribution by presenting a new strategy for waste management (EC, 1989). This strategy aims at enhancing waste prevention and recycling, changing the present hierarchy of disposal options for remaining waste, and preventing massive waste transportation. Various instruments are proposed to reach these objectives, such as harmonization of environmental standards for incineration and landfilling and a series of specific instruments regarding certain products or waste streams.

This paper is concerned with the possible contribution of market-oriented instruments to the general objectives of the EU waste policy. In the next section the Union strategy is briefly confronted with the current situation in the EU. It is argued that the use of product-oriented and other specific instruments may be effective for some waste streams, but will not be sufficient to achieve the overall policy objectives. It also turns out that the proposed harmonization of minimum environmental standards for landfilling and incineration of waste may have substantial side effects on some of these goals. In Section 3 we explore to what extent harmonization of technical standards might have effects on prevention and recycling rates, the disposal hierarchy and waste transportation. In Sections 4 and 5 some suggestions for strengthening the Union strategy on waste management by application of economic instruments are discussed. The EU strategy supports the use of these instruments in general terms, but no concrete proposals have been developed as yet. We focus on discouraging the practice of landfilling, the least preferred disposal option according to the EU waste strategy, by introducing a regulatory charge (Section 4) or a system of tradeable landfill certificates (Section 5). Both economic instruments will function in addition to the harmonization of standards, as discussed in Section 3. Finally, the last section presents a provisional evaluation of both economic instruments.

Policy Objectives and Current Situation in the EU

Recently the European Union set down its objectives concerning waste policy. In May 1990 the Council of Ministers endorsed a 'Community Strategy for Waste Management', which in 1992 became part of the Fifth Environmental Action Programme of the EU, titled *Towards Sustainability.*

In these policy documents, waste is regarded as a (potential) source of pollution, as well as a secondary natural resource. The proposed strategy is to establish a new hierarchy of waste management options on a Union-wide basis. Priority should be given to waste prevention, followed by promotion of recycling and re-use, and then by optimization of final disposal methods for waste which is not re-used. Figure 5.1 gives a simplified picture of the present and the desired structure of waste disposal.

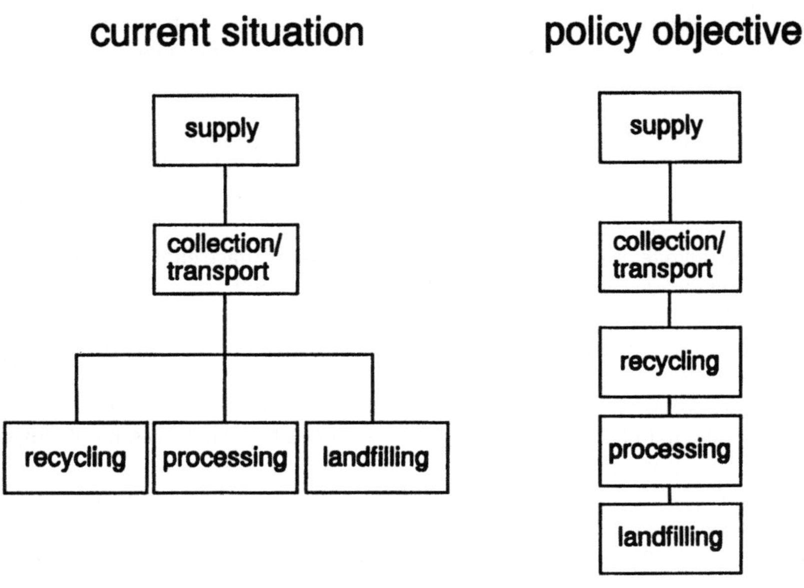

Figure 5.1: Hierarchies of waste management options.

In the current situation, the choice between prevention or supply of waste, and between various options of disposal depends on the available alternatives and their costs. Following the EU objectives, priority should be given to reducing the waste supply by prevention (including in-plant recycling), and to minimizing final disposal by different methods of recovery. Recovery methods include (in order of importance) re-use of products, recycling of material and energy recovery. The last method is not clearly separated from incineration. The EU considers incineration as a method of final disposal, together with the least preferred option, landfilling. In our schedule, processing includes energy recovery and incineration; recycling comprises all other forms of recovery (except in-plant recycling).

One specific aspect concerns transport of waste. The EU proposes the application of a 'self-sufficiency principle' and a 'proximity principle' to avoid massive waste movements, while maintaining open borders between Member States. According to these principles there should be no regional imbalance in the capacity for final disposal. Waste should be treated as close to its place of origin as possible (which for specific types of waste may mean transportation over long distances). Recycling and re-use of waste is exempted from these principles, however.

The Union Strategy for Waste Management is concerned with total waste supply in the Union, estimated at about 2.2 billion tons per annum. This includes household (municipal), industrial and chemical (hazardous) waste.[1] The section on waste management of the Fifth Action Programme limits itself to municipal and hazardous waste. In the Fifth Action Programme a quantitative target for municipal waste can be found. According to this target the supply of municipal waste in the year 2000 should not exceed 300 kg/capita in any Member State, which was the average EU level in 1985. This would amount to a total of 100 million tons, produced by the expected 330 million EU inhabitants in the year 2000 (excluding the eastern part of Germany).

It is likely that realization of the disposal hierarchy and other objectives of the EU waste policy will encounter substantial difficulties, due to the tremendous complexity of the waste market and to the extreme differences in costs of various options both within and between countries. Confining ourselves to municipal and industrial waste, the characteristics are as follows. The collection of municipal waste is usually managed by municipalities, though privatization is taking place. Transport is mainly private. Processing and landfilling in southern Member States are mainly private. In some northern EU countries most landfills are owned by municipalities (e.g. the Netherlands), while in others national and regional utilities are dominant (e.g. Denmark, France). The disposal of industrial and chemical waste is mainly private, and more or less strictly regulated. Recycling is almost exclusively private and only minimally regulated. In addition, usually a mix of local, regional and national regulations is in force.

Table 5.1 provides information about the present costs of landfilling and processing, regarding municipal and industrial waste. The costs of collection of municipal waste depend largely on frequency, e.g. collection once or twice a week, with a bottom line of about 30 ECU/ton. The costs of collection of industrial waste are lower. The costs of prevention and recycling differ for each sub-category of waste and are hard to summarize. In general, in as far as recycling and prevention already exist, their costs are lower than those of landfilling and processing.

According to Table 5.1, processing is much more costly than landfilling. However, there are important differences between facilities and countries. The highest cost of landfilling can be up to 20 times greater than the cost of the lowest one. The cost differential for landfilling is more than 70 ECU/ton of municipal waste. Cost differentials up to a factor of 50 between landfilling and processing may occur. These enormous differences make it attractive to look for cheap landfilling facilities either in the neighborhood or abroad. The costs of transport are lower than the cost differentials between landfilling

Table 5.1: Current costs of disposal methods in EU countries, ECU/ton.

Country	*Processing*			*Landfilling*		
	minimum	maximum	average	minimum	maximum	average
Belgium	27	41	34	19	32	24
Denmark			38	30	43	37
Germany	49	122	85	15	73	44
Greece			30*	2	3	3
France			36			14
Ireland			30*			4*
Italy			30*			4*
Luxemburg			43*			30*
Netherlands	26	56	30	9	30	17
Portugal	25	50	38	6	16	11
Spain	30	30	30	7	10	8
United Kingdom	17	35	26	7	11	9

Source: DHV, 1991. Figures with asterisk are own estimaties

facilities and between landfilling and processing facilities, because there is only a slight increase in the costs of bulk transport per unit of waste over longer distances. Western Europe has an excellent infrastructure for cheap bulk transport by inland waterways. The cost differentials are large enough to make export of waste profitable from e.g. Belgium, Denmark, Germany or Luxembourg to any other country in the EU.

The development of an EU waste policy which is able to cope with these difficulties is still in its infancy. According to the Fifth Environmental Action Programme a variety of actions will be taken during the present decade. They can be divided into five groups:
- Actions to stimulate prevention and recovery on a case-by-case basis, e.g. support for cleaner technologies and better product design.
- Announcement of new regulations regarding specific waste streams (like packaging, waste oils, batteries), specific environmental aspects (like dioxin emissions) and certain disposal options (landfilling and incineration).
- Actions intended to have a general effect, like the promotion of waste management plans in all Member States, the introduction of a system of civil liability for environmental damage, the opening of markets for recycled materials and the provision of reliable data.
- An integrated approach for various priority waste streams, such as used tires and CFCs, primarily based on voluntary agreements with the industries involved.
- General support concerning the use of economic incentives and instruments.

To a large extent, the proposed EU strategy is directed towards specific products and waste streams. Apart from its relevance for other objectives (promotion of cleaner technologies), it is feared that this type of action is insufficient to reach the general objectives of changing the disposal hierarchy and minimizing waste transports. For this reason the product-specific policies are not dealt with in this paper.

Other elements of the EU strategy may have a noticeable impact on the disposal hierarchy and transport of waste, particularly the announced standards for final disposal options. Some of these elements have already been translated into new directives (for example the framework directive on waste 75/442, renewed in 1991) or are currently being decided on (for example the directive on landfilling); others are still in the early

Table 5.2: Situation in 1990 regarding municipal and industrial waste.

	Municipal waste, mln tons, situation in 1990			costs (mln ECUs) of landfill and processing	Industrial waste, mln tons, situation in 1990			costs (mln ECUs) of landfill and processing
	landfill	proces-sing	recycling		landfill	proces-sing	recycling	
Belgium	1.8	1.2	0.4	85	54.1	35.8	11.2	2481
Denmark	0.8	1.6	0.2	90	2.2	4.8	0.5	262
Germany	21.6	5.6	0.8	1423	165.7	49.2	32.1	11471
Greece	2.5	0.2	0.3	13	13.1	0.9	1.5	67
France	9.7	8.1	0.7	427	312.8	242.9	0.7	13124
Ireland	1.1	0.1	0.1	7	27.5	1.9	1.4	174
Italy	17.8	1.3	0.9	114	136.2	9.6	7.2	868
Luxemburg	0.0	0.1	0.0	6	0.9	4.1	0.8	203
Netherlands	4.3	2.8	0.3	157	36.0	21.7	0.3	1263
Portugal	1.9	0.3	0.3	33	0.2	0.0	0.0	3
Spain	7.5	3.2	1.9	155	135.7	57.2	33.4	2800
United Kingdom	11.5	5.0	3.5	233	185.5	80.7	56.7	3768
Total	80.6	29.4	9.4	2742	1069.8	507.8	145.7	36485

Source: OECD (1989 and 1991) and Eurostat (1991); cost calculations are based on tabel 5.1.

stages of preparation. This makes it difficult to analyze the extent to which enforcement of these new regulations will help achieve the general goal of the EU waste policy. Nevertheless, we will explore this question in the next section. As a starting point Table 5.2 provides some data on the present amounts and disposal costs of municipal and industrial waste.

Regulation of Final Disposal Methods

Harmonization of technical standards for final disposal of waste is an important element of the EU waste policy. In June 1989 two directives on the incineration of municipal waste by new (89/369) and existing installations (89/429) came into force. A directive on incineration of industrial (non-hazardous) waste is still in preparation. In May 1991 the Commission proposed a new directive on landfilling of waste, covering municipal and industrial waste, as well as existing and new landfills.

These directives have common aims: harmonization of emission standards and other environmental conditions, establishing a minimum level of environmental care throughout the Union, and harmonization of tariffs. The undisturbed functioning of the Internal Market is one motive behind the latter, but two other motives seem to be even more important. Tariff harmonization is also seen as a remedy for massive waste transports, while higher tariffs for incineration and landfill provide a stimulus for waste prevention and re-use. The directive on landfill in particular, emphasizes these motives and stipulates that tariffs must be sufficient to cover at least all costs of construction, exploitation, closure and post-closure care.

There is no doubt that implementation of the new directives will lead to higher costs of landfilling and incineration in those Member States where existing standards are less stringent. If the higher costs are to be paid by the waste suppliers, this will also create a stimulus for waste prevention and recycling, depending on the marginal costs of these alternatives.

Stricter environmental standards could also influence the present cost differences between Member States, but the extent and even the direction of this effect is uncertain. If proportional cost increases in all countries are more or less comparable, absolute differences will increase. If cost increases are relatively substantial in countries with low technical standards, differences between countries can decrease. However, equalization of costs is clearly not a realistic expectation, because of remaining differences between environmental standards as well as enduring structural differences between Member States.[2]

To estimate the possible effects of the new EU directives on the disposal hierarchy and waste transports in the year 2000, some projections have been made. The results for municipal and industrial waste in the EU as a whole are shown in Table 5.3. Computations for each of the Member States are reproduced in the Appendix (Tables A.1 - A.8). In all projections it is assumed that the EU legislation is fully implemented and that suppliers of waste pay the full cost of each method of disposal. Calculations are made in comparison with the expected supply of waste for landfilling plus incineration. This expected supply is extrapolated from actual data, assuming a 2% annual increase in waste volumes. The expected volume of municipal waste in the year 2000 is 146 million tons. The total volume of industrial waste will be 2100 million tons.

Table 5.3: Effects of regulation on waste management options and costs in the EU in the year 2000.

	Volumes (mln ton)					Costs (mln ECU)			
	Trade	Recycling	Landfill	Proces.	Total	Recycling	Landfill	Proces.	Total
DIVERGING RATES									
Municipal waste									
- no trade	-	47	81	21	146	736	1829	1232	3797
- with trade	14.0 (24.8)	40	87	23	146	591	1615	1251	3458
Industrial waste									
- no trade	-	550	1247	379	2101	9998	26225	20571	56794
- with trade	239.6 (365.6)	489	1286	408	2101	8315	22779	20989	52083
CONVERGING RATES									
Municipal waste									
- no trade	-	63	68	18	146	1224	3019	1698	5940
- with trade	9.2 (13.5)	57	73	19	146	1065	3125	1775	5965
Industrial waste									
- no trade	-	956	897	310	2101	24268	39741	28065	92073
- with trade	122.0 (169.2)	889	951	326	2101	22268	40628	29146	92042

In the first projection, it is assumed that stricter environmental standards result in a 4% annual increase in average landfill and incineration costs[3] in all countries ('diverging rates'). In the second projection, the same annual increase is assumed in Germany together with a faster increase in other Member States, with the result that by the year 2000, 50% of the initial (relative) cost differences compared with Germany will have disappeared ('converging rates'). The projected level of costs in Germany is taken as a hallmark because this is where most progress has been made in implementing strict environmental standards.

In this way, two possible outcomes of the harmonization of environmental standards are modelled.

The higher costs of landfilling and incineration induce more recycling and prevention in both projections.[4] For convenience, we assume that recycling and prevention rates increase until their estimated marginal costs equal the costs of landfilling and incineration. Clearly, this is a very optimistic assumption because there are several bottlenecks in the logistics of prevention and recycling, in sales of recycled materials, etc. Therefore, the presented estimates of recycling and prevention should be considered as the potentially feasible maxima.

Both projections have two variants. In the first variant no transport of waste between countries takes place, while in the second variant the possibility of trading is introduced. The real trade volume is hard to predict because the capacity of low-cost landfills and incinerators is unknown. If there is enough capacity, the marginal benefits of exporting waste outweigh the marginal costs of transport and virtually all waste can be exported to countries like Greece, Italy and Ireland. It is probable, however, that low-cost capacity is not unlimited and that certain restrictions will be introduced to limit waste inflow and outflow. Therefore, it is assumed that trade proceeds proportionally with the average cost differential between countries but with a maximum of 25% of the total inland waste volume of the exporting country. No limit is set on the capacity of the importing country. For this reason, the estimates of trade volumes should also be considered as potentially feasible maxima.

The costs of landfilling and processing are calculated per country using the amounts of waste handled. This implies that the exporting countries show costs that are too low and importing countries show costs that are too high because in practice the exporting country will pay the importing country. The overall costs, however, are correct. It is furthermore assumed that imported waste will not be recycled. The total amount of waste for landfilling, after all other calculations, has been increased by 20% (the ashes) of the amount of waste for processing. Therefore the amounts do not add up to the total amount of waste.

Results with Diverging Rates of Landfilling and Processing

With a 4% annual increase in landfill and incineration costs and assuming no trade possibilities, recycling (including prevention) of municipal waste will increase to a volume of 47 million tons, which is about 32% of the total waste. This is a substantial increase compared with the percentage in 1990 (approximately 7.5%). Total costs of recycling, landfilling and processing municipal waste in the year 2000 will be 3.8 billion ECUs.

If trade is introduced between the EU countries, around 25 million tons of municipal waste will be transported. The nett trade (that is the balance of imports and exports of all country) is 14 million tons. The tables in the Appendix show that Germany and France are the biggest exporters; the major receivers are Greece and Ireland. The volume of recycling drops to 40 million tons, which is about 27% of the total waste. Total costs of recycling, landfilling and processing waste will be approximately 10% lower due to the transport of waste: 3.5 billion ECUs. The savings from trade are about 300 million ECUs (excluding transportation costs).

The results for industrial waste resemble those for municipal waste. Without trade, the total amount of recycling is about 550 million tons (26%, compared to 8.5% in 1990). The total costs of industrial waste treatment in 2000 are estimated at 57 billion ECUs. If trade is introduced, around 366 million tons will be transported, which is more than 17% of total industrial waste. The nett trade balance is some 240 million tons. France and Germany are again the biggest exporters; Greece, Ireland and Italy the major receivers. In comparison with the 'no-trade variant', recycling decreases by more than 10% to a total of 490 million tons. In this scenario the total costs for industrial waste treatment are 52 billion ECUs. The savings from trade are 4700 million ECUs.

Results with Converging Rates of Landfilling and Processing

If the costs of landfilling and processing in the EU Member States converge to the (high) German level, recycling of waste is further stimulated. If there is no trade assumed, the total level of recycling of municipal waste in the year 2000 is 63 million tons, which is 43% of the total amount. The total costs are much higher than in the previous scenario, because of the higher costs of landfilling and processing in all Member States. Total costs amount to 5.9 billion ECUs.

With the assumption of trade of waste between countries, less recycling is done: 57 million tons (39%). Total costs however are about the same, because trade mainly replaces recycling. The amount of municipal waste transported is 13.5 million tons (9% of total waste volume). Nett transport is over 9 million tons.

In the case of industrial waste, with no trade and rates converging to the German level, the total amount of recycling is 956 million tons, which is 45% of total industrial waste produced. Total costs are again much higher than in the first scenario: 92 billion ECUs.

If trade is accepted, less recycling is to be expected: 889 million tons. The amount of waste transported is almost 170 million tons (8% of the total). Nett transport is 122 million tons. Total costs again are 92 billion ECUs.

Overall Evaluation

From these computations some conclusions may be drawn regarding the general goals of the EU waste policy. First of all, it seems likely that enforcement of EU regulations on landfilling and incineration will have a strong positive influence on the level of recycling and prevention. Keeping in mind that the estimates concerning recycling and prevention in Table 5.3 represent potentially feasible maxima, these figures nevertheless point to an important positive side effect of the harmonization of emission standards and other environmental conditions of final disposal methods in the EU. Secondly, the harmonization approach fails to change the present hierarchy of disposal options. On the contrary, all scenarios show a relative increase in the amount of waste for landfilling (the least preferred option) compared to the amount for processing. The last conclusion may be that, given the scenario assumptions, there are extensive opportunities for cost-saving transportation of waste. The harmonization of environmental standards is ineffective in discouraging this option.

Regulatory Charges on Landfilling Waste

The results of the last section show that higher costs of incineration and landfilling will stimulate prevention and recycling to a certain extent. However, the pursued priority change in the disposal hierarchy will not be reached, while massive waste transportation will arise if trade possibilities exist. The use of regulatory charges is often proposed as an appropriate market-oriented instrument to strengthen the shift towards prevention and recycling. On a national or regional scale some countries have already introduced regulatory charges on landfilling and/or incineration or have announced their intention to do so (Denmark, Belgium, Germany, the Netherlands). Apparently, these additional charges are implemented in the countries with relatively high landfill and incineration costs. Therefore, besides their effects on recycling and prevention, they will stimulate waste transportation if trade is allowed.

In this section we explore the possible effects of a regulatory charge, introduced at the EU level. Its objective would be to enhance prevention and recycling, while at the same time pushing back or at least not encouraging waste transportation. This charge may also be useful as an instrument to change the present disposal hierarchy. To examine its possible effectiveness in this respect, we assume that the incentive charge is imposed only on waste, offered for landfilling. Furthermore we assume that the incentive charge will be imposed on top of the charge with a financing objective (diverging rates) as discussed in the last section.

A European waste charge with regulatory purposes could in principle be designed as an *ad valorem* charge, a specific charge or a cost-equalizing charge. Introducing an *ad valorem* charge (raising the costs of landfilling in all Member States by a certain, uniform percentage) could stimulate prevention and recycling, but would clearly enlarge the attractiveness of waste transportation. For this reason the *ad valorem* charge is not a viable option. This type of regulatory charge simply strengthens the effects of higher, diverging rates as discussed in the previous section. A specific charge (raising the costs of landfilling in all Member States by a uniform amount per ton) reduces the relative cost differences between countries. A cost-equalizing charge would lead to uniform costs of landfilling in all Member States. In principle these two designs represent tenable options.

In our computations the specific charge is set at a rate of 16 ECUs per ton. This tariff is equivalent to 25% of the projected landfill costs in Germany, namely 65 ECU/ton in the year 2000. The equalizing charge will result in a uniform price of landfilling in all Member States of 65 ECU/ton in the year 2000. The effects of both incentive charges are computed for situations with and without trade possibilities. The assumptions regarding switching between disposal options and recycling, and regarding trade volumes are the same as those in the previous section. An additional assumption concerns switching between landfilling and processing. It is assumed that as soon as the landfill costs become higher than the costs of processing, half of the potential amount of landfilled waste will be processed instead (the costs of processing increase by 4% annually, according to the diverging rates scenarios in the previous section). Table 5.4 shows the effects of both regulatory charges on different disposal options and on waste transport for the EU as a whole (in comparison with the effects of diverging rates according to the last section). Computations for each of the Member States are reproduced in the Appendix (Tables B.1 - B.8).

Results with a Specific Incentive Charge on Landfilling

The results of the calculations show that a specific charge on landfilling will stimulate recycling compared with the diverging rates scenarios. The recycling rates for municipal waste are 36% (no trade) and 31% (with trade) respectively.

The level of trade decreases somewhat, due to the effect of lower cost differentials. The total amount of trade is about 21 million tons (the nett trade volume is 12.5 million tons). Total costs are, of course, higher because of the charge (5 billion ECUs without trade and 4.8 billion ECUs with trade respectively). In some countries the landfill costs will equal the processing costs and shifts from landfilling to processing can be expected. The main shift, however, is between landfilling and recycling.

For industrial waste the results are comparable to the results for municipal waste. Total recycling increases to 38% (no trade) and 33% (with trade). When trade is introduced

Tabel 5.4: Effects of regulatory charges on waste management options and costs in the EU in the year 2000.

	Volumes (mln ton)					Costs (mln ECU)			
	Trade	Recycling	Landfill	Proces.	Total	Recycling	Landfill	Proces.	Total
DIVERGING RATES (Table 5.3)									
Municipal waste									
- no trade	-	47	81	21	146	736	1829	1232	3797
- with trade	14.0 (24.8)	40	87	23	146	591	1615	1251	3458
Industrial waste									
- no trade	-	550	1247	379	2101	9998	26225	20571	56794
- with trade	239.6 (365.6)	489	1286	408	2101	8315	22779	20989	52083
SPECIFIC CHARGE									
Municipal waste									
- no trade	-	52	76	22	146	896	2867	1278	5041
- with trade	12.25 (20.8)	45	82	24	146	715	2807	1286	4808
Industrial waste									
- no trade	-	797	984	400	2101	17337	36129	21645	75110
- with trade	211.3 (315.0)	683	1079	424	2101	14022	35807	21794	71624
EQUALIZING CHARGE									
Municipal waste									
- no trade	-	73	41	40	146	1508	2672	2101	6281
- with trade	3.1 (4.0)	71	40	43	146	1457	2632	2270	6359
Industrial waste									
- no trade	-	907	611	728	2101	19857	39816	37658	97331
- with trade	58.2 (68.9)	881	624	744	2101	19067	40650	36550	96267

about 315 million tons of waste will be transported, which is 14% less compared with the diverging rates scenario (nett transport is about 211 million tons). Total costs rise to 75 billion ECUs (without trade) or 72 billion ECUs (with trade). Savings from trading industrial waste are 3500 million ECUs.

Results with a Cost-Equalizing Charge on Landfilling

If a cost-equalizing charge on landfilling is introduced, the recycling rate for municipal waste rises to (almost) 50% in both projections. These scenarios also show a substantial shift from landfilling towards processing, because in several countries landfill tariffs will be higher than the tariffs of processing. The amount of trade is very small (it may only be interesting for waste to be processed): only 4 million tons (the nett trade volume is 3.1 million tons). Total costs are higher because of the cost-equalizing charge and the changes in handling methods, and are 6.3 billion ECUs (without trade) and 6.4 billion ECUs (with trade) respectively.

The results for industrial waste are comparable. The recycling rates for industrial waste are slightly above 40%, while processing increases to 35%. Trading of industrial waste decreases to approximately 3%. Again total costs are higher, ending up with almost 100 billion ECUs in both cases (with and without trade).

Overall Evaluation

Generally speaking, both regulatory charges result in more prevention and recycling, in a shift towards incineration (especially in those countries with low landfill costs to date), and in less waste trade. The specific charge leads to positive but relatively modest results with regard to the goals of the EU waste policy. The cost-equalizing charge is particularly effective in reducing the amount of waste for landfilling and waste transports.

However, there are also drawbacks. Firstly, introducing the specific charge will not be sufficient to achieve the policy objectives regarding the disposal hierarchy, and its contribution to the other EU-policy goals is rather limited. Secondly, the rather large cost increase associated with the specific charge and the even larger cost increase in the case of the cost-equalizing charge, makes their implementation rather difficult. Despite the fact that these cost increases are largely made up of extra government revenues (which can be used for additional public activities or be compensated by tax reductions), it seems improbable that, especially the EU countries with low-cost landfills, would endorse the introduction of these charges. A third problem may be that the effectiveness of these charges is doubtful in the light of possibilities for fraud and illegal dumping of waste. Because of these drawbacks we have to conclude that the implementation of a specific or cost-equalizing regulatory charge is not very feasible. Moreover, with the specific charge, the policy objectives are only partly achieved.

Tradeable Landfill Certificates

Tradeable permits are often presented as a policy instrument which combines a clear-cut policy goal with the advantages of using the market mechanism. Application of this instrument may be attractive to reach the goals of changing the disposal hierarchy and preventing waste transports in an effective and efficient way, together with enforcement of minimal environmental standards as discussed earlier.

Tradeable permits could be given the form of site-specific 'landfill certificates' or 'disposal permits', giving the right to dispose of a certain amount of waste on that landfill site. Initially these certificates would be issued by the environmental authority. As always they can be grandfathered, for example to the landfill owners, or auctioned to landfill owners and waste suppliers.

The certificates should match the residual capacity of each landfill site. In all, they should also match the total amount of landfilling the authority is prepared to allow. The certificates would be tradeable between landfill owners and waste suppliers. Waste suppliers could buy certificates when they deliver waste for disposal (if the landfill owner still has certificates in stock) or in advance, ensuring their future disposal possibilities. Certificates could also be traded between waste suppliers.

The certificates would expire when the corresponding amount of waste is actually offered and accepted for disposal. The control of disposal is done by the environmental authority on the basis of the quantity of expired site capacity. Site-specific certificates and a periodical measurement of the remaining landfill capacity are needed. Also illegal landfills must be effectively controlled.

The element of scarcity of landfill capacity introduced in this system, will lead to a positive price of the landfill certificates. For waste suppliers, total and marginal costs of landfilling will rise, thus providing an incentive to reduce their demand for this kind of waste disposal. For landfill owners, selling landfill certificates creates a source of extra profits, assuming that the certificates are initially grandfathered. If they are auctioned initially, these profits are skimmed off by the environmental authority. In this case landfill owners would only have to deal with secondary changes in prices and quantities, with probably a more or less neutral effect on profits.

Remembering that the costs of disposal on various landfill sites will differ, even after harmonization of minimal environmental standards (see above), certificates for cheaper landfill sites will be relatively attractive as long as transport costs are not prohibitive. As a result, prices of the (site-specific) certificates differ in a complementary way to the costs of disposal on each site and the costs of transport to each site. In other words, total costs of landfilling per ton of waste, including the costs of certificates, transport and disposal, tend to equalize, independent of the location of waste supply and landfill site. This result resembles that of the cost-equalizing charge (see the preceding section), with one important difference. In the system with tradeable certificates transport costs are included in the equalization process (implying that transporting waste will still be attractive[5]); in the case of the equalization charge they are not. The effects on recycling will be very similar to the cost-equalizing charge. Total costs may be lowér because of savings of exporting waste to the countries with low-costs of landfilling.

Another major advantage of this instrument in comparison with the cost-equalizing charge is that the proceeds from secondary sales of certificates (i.e. capacity of landfills) remain in the same hands, while the charge must be transferred to a fund or any authority, which creates additional administrative inefficiencies.

This short outline of a policy based on the instrument of tradeable landfill certificates can be refined in several ways. The instrument of tradeable landfill certificates could, for example, also be used to shorten the transition period towards harmonized environmental standards. Usually, the lower the standards, the lower the disposal costs for the supplier appear. Waste suppliers will try to get certificates for landfill sites with lower standards first, and when they are no longer available or when total costs of landfilling have become equal for all sites, they will turn to certificates for facilities with higher standards. So, the landfill sites with low quality will be utilized relatively fast. The transition to facilities which at least meet harmonized environmental standards can be accelerated by authorities buying out certificates for low quality sites. In this way the instrument of tradeable certificates offers an alternative to enforcement of harmonized standards by command-and-control instruments.

A particularly interesting feature of the application of tradeable landfill certificates to alter the present disposal hierarchy is the creation of a new market and new market interests. This could help to limit problems of fraud, such as illegal dumping of waste not suitable for specific landfill sites. At present, waste suppliers are in principle only interested in disposing of their waste at the lowest cost, while landfill owners are interested in a profitable exploitation of their existing facilities. Both participants can use legal or illegal means to reach their goals. This leaves the public authorities with the responsibility for the enforcement of (environmental) rules and for guaranteeing future disposal capacity. This division of tasks and interests does not change by introducing stricter environmental rules or regulatory charges. Consequently, fraud by suppliers, landfill owners or authorities becomes more attractive with higher costs of landfilling and might only be restricted by extensive and complex control and enforcement schedules. Illegal disposal in one way or another may indeed become a fully rational choice.

The introduction of tradeable certificates creates a potentially important new market interest. Certificates represent a right to future disposal at certain landfill sites. This means that holders of certificates will have an interest in the future capacity and quality of the facility. Acceptance of waste from non-certificate holders for example, limits the capacity for future disposal. Acceptance of unsuitable types of waste reduces the quality of the site and can be translated into capacity loss and therefore losses of future proceeds.

On the whole, non-fraudulent behavior becomes a condition for certificate holders (suppliers and landfill owners) to earn back their investment in certificates. This will probably not eliminate illegal behavior totally, but it may well keep it at a manageable level. Control need no longer be based on individual waste transports, but can be based on periodical measurement of the quality and remaining capacity of the landfill sites.

The main conclusion is that tradeable certificates can be an effective instrument to limit landfilling to politically chosen quantities, which is a crucial element in changing the disposal hierarchy. Also, the most efficient alternatives are stimulated. However, this instrument is not a barrier to waste transportation. On the contrary, it may initially encourage more transports. If trade is not allowed, the effects of the instrument are restricted to the national level. Substantial cost increases can be avoided if the certificates are grandfathered initially. Moreover, the introduction of a system of tradeable landfill certificates creates an incentive for non-fraudulent behavior, which will help to reduce illegal disposal and fraud.

Conclusions

In light of the central objectives of the EU waste policy - priority for waste prevention and recycling, preference for processing above landfilling, and minimizing waste transportation -, there is a need for a general instrument in solid waste policy. The announced standards of the EU for final disposal options will probably lead to higher costs of waste disposal. Although this will have positive effects on prevention and recycling, the proposed hierarchy of disposal options will not be reached and massive waste transports to low-cost countries will take place.

The imposition of regulatory charges on waste for landfilling is conducive to the policy objectives. An *ad valorem* charge, however, increases the cost differential between the EU countries. This will encourage trade of waste between countries and will have little effect on recycling. A specific charge will somewhat decrease the (relative) cost differences between the EU countries and will therefore lead to higher recycling rates and somewhat less trade. These effects are not significant enough to achieve the desired change in the disposal hierarchy. A cost-equalizing charge will probably result in strongly diminishing trade and high recycling rates. Some extra processing is also expected. This type of regulatory charge however would also lead to very high extra costs for waste management. The last two types of regulatory charges would also create a need for additional controls on waste handling in the EU. Apart from the resulting cost increase, the risk of fraud and illegal dumping is a serious drawback of these charges.

Another possibility is the introduction of tradeable landfill certificates. With this instrument the total landfill capacity can be limited, which is a crucial element in changing the disposal hierarchy. The possibility of trade however is not affected. In short, the effectivity is comparable to the cost-equalizing charge, but with higher efficiency. The landfill certificates create a new market interest for their holders (suppliers of waste and landfill owners). Although fraud and illegal dumping practices will not be completely eliminated, they can presumably be kept at manageable levels.

Notes

[1] It is very difficult to estimate the total waste supply produced in the Union, because of different definitions used by Member States and lack of reliable data. Later in this paper we refer to OECD data regarding present amounts of municipal and industrial waste (cf. Table 5.2).

[2] In some Member States actual environmental standards are already higher than prescribed by the new EU directives. One relevant structural factor is the availability of space for landfilling.

[3] This is based on trends in the Netherlands.

[4] Usually prevention is invisible in ex post waste statistics (e.g. Table 5.2). In the following prognostic tables prevention is included in recycling. Marginal costs of recycling and prevention are based on current figures for specific waste components in the Netherlands.

[5] If trade is not allowed, equalization of landfill costs would only occur in each Member State. In this case the effects of the instrument must be evaluated on a country-by-country basis.

References

Bohm, P. and C.S. Russel (1985), 'Alternative Policy Instruments'. In A.V. Kneese and J.L. Sweeney (eds), *Handbook of Natural Resources and Environmental Economics,* Volume 1, North Holland, Amsterdam.

Brown, G. jr. (1984), *Selected Economic Policies for Managing Hazardous Waste in Western Europe,* Department of Economics, University of Washington, Seattle.

DHV (Dwars, Hederik en Verhey) (1991), *Afvalstoffenbeleid in internationale context,* Publikatiereeks afvalstoffen, Nr. 1991/5g, Dutch Ministry for Housing, Regional Development and the Environment, The Hague.

EC (European Community) (1989), *Community Strategy for Waste Management,* SEC (89) final, Brussels.

EC (European Community) (1992), *Towards Sustainability (Fifth Environmental Action Programme),* Com (92) 23 final.

Eurostat (1991), *Environment Statistics 1991,* Eurostat, Luxembourg.

OECD (Organization for Economic Cooperation and Development) (1989/91), *Environmental Data Compendium 1989 and 1991,* OECD, Paris.

Pearce, D.W. and R.K. Turner (1992), *Market-based Approaches to Solid Waste Management,* CSERGE working paper Nr. WM 92-02, University of East Anglia, Norwich.

Russel, C.S. (1988), 'Economic Incentives in the Management of Hazardous Wastes', *Columbia Journal of Environmental Law,* Vol. 13, 1988, pp. 257-274.

Appendix - Computation Results for the EU Member States

A - Projections of the effects of regulation of final disposal methods; diverging and converging rates

Table A.1: Diverging rates, municipal waste, no trade

	Volumes (mln ton)				Costs (mln ECU)			
	Recycling	Landfill	Processing	Total	Recycling	Landfill	Processing	Total
Belgium	1	2	1	4	16	77	43	136
Denmark	2	1	1	3	34	38	59	132
Germany	18	14	3	34	437	889	392	1718
Greece	1	3	0	4	3	13	6	21
France	8	11	5	23	132	222	266	620
Ireland	0	1	0	2	2	7	2	11
Italy	5	19	1	24	24	121	36	181
Luxemburg	0	0	0	0	2	2	6	9
Netherlands	3	5	2	9	44	118	79	241
Portugal	1	2	0	3	4	36	12	53
Spain	4	9	2	15	36	109	108	253
United Kingdom	5	15	6	24	3	197	222	422
Total	47	81	21	146	736	1829	1232	3797

Table A.2: Diverging rates, municipal waste, with trade

	Volumes (mln ton)					Costs (mln ECU)			
	Trade	Recycling	Landfill	Proces.	Total	Recycling	Landfill	Proces.	Total
Belgium	-0.1	1	2	1	4	14	75	51	139
Denmark	-0.3	1	1	1	3	28	39	58	125
Germany	-8.3	14	10	2	34	328	667	294	1289
Greece	7.5	1	10	1	4	3	45	25	73
France	-3.5	7	9	4	23	112	184	236	532
Ireland	4.0	0	5	0	2	1	30	22	54
Italy	1.5	4	21	1	24	22	130	55	208
Luxemburg	0.4	0	0	0	0	1	15	12	29
Netherlands	-0.4	3	4	2	9	41	108	96	246
Portugal	0.6	1	3	0	3	4	45	18	67
Spain	-0.3	4	9	3	15	35	104	125	263
United Kingdom	-1.2	5	13	7	24	3	172	259	434
Total	14.0 (24.8)	40	87	23	146	591	1615	1251	3458

Table A.3: Diverging rates, industrial waste, no trade

	Volumes (mln ton)				Costs (mln ECU)			
	Recycling	Landfill	Processing	Total	Recycling	Landfill	Processing	Total
Belgium	53	48	26	122	1026	1701	1319	4045
Denmark	4	2	4	9	87	130	202	419
Germany	149	130	27	301	3498	8495	3454	15447
Greece	2	16	1	19	10	72	33	114
France	124	407	183	678	3132	8441	9758	21331
Ireland	2	34	2	37	21	213	67	301
Italy	13	168	8	187	105	1054	337	1497
Luxemburg	3	1	3	7	61	60	195	317
Netherlands	19	38	17	71	448	950	760	2157
Portugal	0	0	0	0	0	2	1	4
Spain	70	170	45	276	668	2010	1999	4677
United Kingdom	110	232	64	394	941	3097	2446	6485
	550	1247	379	2101	9998	26225	20571	56794

Table A.4: Diverging rates, industrial waste, with trade

	Volumes (mln ton)					Costs (mln ECU)			
	Trade	Recycling	Landfill	Proces.	Total	Recycling	Landfill	Proces.	Total
Belgium	-14.7	45	40	27	122	827	1438	1343	3608
Denmark	0.8	3	3	4	9	71	175	242	488
Germany	-65.5	121	98	21	301	2623	6371	2591	11585
Greece	104.4	2	114	9	19	10	505	413	928
France	-130.5	105	316	158	678	2665	6553	8409	17628
Ireland	65.3	2	92	10	37	20	581	446	1047
Italy	49.4	13	210	16	187	102	1324	707	2133
Luxemburg	1.9	3	3	3	7	47	154	221	422
Netherlands	2.0	17	36	25	71	387	906	1116	2410
Portugal	15.7	0	15	1	0	0	242	73	315
Spain	-9.4	68	157	52	276	636	1856	2314	4806
United Kingdom	-19.5	109	201	81	394	927	2674	3114	6714
Total	239.6 (365.6)	489	1286	408	2101	8315	22779	20989	52083

Table A.5: Converging rates, municipal waste, no trade

	Volumes (mln ton)				Costs (mln ECU)			
	Recycling	Landfill	Processing	Total	Recycling	Landfill	Processing	Total
Belgium	2	1	1	4	41	75	69	185
Denmark	2	1	1	3	39	41	90	169
Germany	18	14	3	34	437	889	392	1718
Greece	1	3	0	4	5	94	10	109
France	12	7	5	23	282	305	415	1002
Ireland	0	1	0	2	3	40	4	46
Italy	6	18	1	24	45	641	62	748
Luxemburg	0	0	0	0	2	2	8	12
Netherlands	5	3	2	9	111	140	136	387
Portugal	1	1	0	3	27	60	19	105
Spain	5	9	2	15	49	342	190	581
United Kingdom	11	10	4	24	183	391	303	877
Total	63	68	18	146	1224	3019	1698	5940

Table A.6: Converging rates, municipal waste, with trade

	Volumes (mln ton)					Costs (mln ECU)			
	Trade	Recycling	Landfill	Proces.	Total	Recycling	Landfill	Proces.	Total
Belgium	0.1	2	2	1	4	36	85	83	203
Denmark	-0.0	1	1	1	3	35	43	98	176
Germany	-8.3	14	10	2	34	328	667	294	1289
Greece	2.5	1	5	0	4	5	174	31	210
France	-0.9	11	7	5	23	261	306	410	977
Ireland	2.0	0	3	0	2	3	105	25	133
Italy	1.8	6	20	1	24	44	700	83	828
Luxemburg	0.4	0	0	0	0	2	19	19	40
Netherlands	0.1	4	3	2	9	103	152	156	411
Portugal	0.8	1	2	0	3	25	93	31	148
Spain	0.8	5	10	2	15	49	369	210	627
United Kingdom	0.6	11	11	4	24	176	412	336	924
Total	9.2 (13.5)	57	73	19	146	1065	3125	1775	5965

Table A.7: Converging rates, industrial waste, no trade

	Volumes (mln ton)				Costs (mln ECU)			
	Recycling	Landfill	Processing	Total	Recycling	Landfill	Processing	Total
Belgium	60	45	21	122	1534	2265	1843	5641
Denmark	5	2	3	9	145	134	260	540
Germany	149	130	27	301	3498	8495	3454	15447
Greece	8	10	1	19	163	363	51	577
France	286	276	146	678	9418	11830	13093	34341
Ireland	15	22	1	37	342	781	104	1227
Italy	73	109	6	187	1698	3875	523	6096
Luxemburg	4	1	2	7	112	66	232	410
Netherlands	29	31	14	71	863	1403	1179	3445
Portugal	0	0	0	0	2	4	1	7
Spain	132	114	36	276	2719	4392	3100	10211
United Kingdom	196	156	51	394	3773	6132	4225	14131
Total	956	897	310	2101	24268	39741	28065	92073

Table A.8: Converging rates, industrial waste, with trade

	Volumes (mln ton)					Costs (mln ECU)			
	Trade	Recycling	Landfill	Proces.	Total	Recycling	Landfill	Proces.	Total
Belgium	-9.3	55	40	22	122	1371	2030	1957	5358
Denmark	1.4	4	3	4	9	132	188	362	682
Germany	-65.5	121	98	21	301	2623	6371	251	11585
Greece	35.1	8	43	4	19	163	1486	354	20003
France	-47.2	264	254	141	678	8783	10891	12643	32317
Ireland	29.0	14	48	5	37	338	1721	407	2466
Italy	27.6	72	134	10	187	1681	4783	822	7286
Luxemburg	2.7	4	3	3	7	102	188	319	608
Netherlands	2.5	26	33	17	71	806	1491	1473	3770
Portugal	11.3	0	10	1	0	2	422	115	539
Spain	7.8	129	123	40	276	2619	4738	3377	10734
United Kingdom	4.7	191	161	58	394	3647	6319	4727	14693
Total	122.0 (169.2)	889	951	326	2101	22268	40628	29146	92042

B - Projections of the effects of incentive charges, imposed on landfilling; specific and cost-equalizing charges

Table B.1: Specific charge of 16 Ecu/ton, municipal waste, no trade

	Volumes (mln ton)				Costs (mln ECU)			
	Recycling	Landfill	Processing	Total	Recycling	Landfill	Processing	Total
Belgium	2	1	2	4	36	50	77	163
Denmark	2	0	1	3	36	34	72	143
Germany	19	13	3	34	497	1031	392	1920
Greece	1	3	0	4	4	56	6	65
France	8	11	5	23	135	389	266	790
Ireland	0	1	0	2	2	25	2	29
Italy	6	18	1	24	35	405	36	475
Luxemburg	0	0	0	0	2	2	6	10
Netherlands	5	3	2	9	101	129	79	309
Portugal	1	2	0	3	5	70	112	87
Spain	4	9	2	15	38	251	108	397
United Kingdom	5	15	6	24	5	426	222	653
Total	52	76	22	146	896	2867	1278	5041

Table B.2: Specific charge of 16 Ecu/ton, municipal waste, with trade

	Volumes (mln ton)					Costs (mln ECU)			
	Trade	Recycling	Landfill	Proces.	Total	Recycling	Landfill	Proces.	Total
Belgium	-0.1	2	1	2	4	29	63	76	168
Denmark	-0.3	1	1	1	3	29	39	67	136
Germany	-8.3	14	10	2	34	373	773	294	1440
Greece	4.8	1	7	1	4	4	146	25	176
France	-3.5	7	9	4	23	114	324	236	674
Ireland	3.5	0	4	0	2	2	95	22	119
Italy	2.8	5	21	1	24	34	462	55	550
Luxemburg	0.4	0	0	0	0	1	21	12	35
Netherlands	-0.4	4	3	2	9	84	128	96	309
Portugal	0.7	1	3	0	3	4	91	18	114
Spain	0.4	4	9	3	15	36	258	125	419
United Kingdom	-0.0	5	14	7	24	4	406	259	669
Total	12.5 (20.8)	45	82	24	146	715	2807	1286	4808

Table B.3: Specific charge of 16 Ecu/ton, industrial waste, no trade

	Volumes (mln ton)				Costs (mln ECU)			
	Recycling	Landfill	Processing	Total	Recycling	Landfill	Processing	Total
Belgium	55	30	47	122	1110	1532	2346	4987
Denmark	4	2	4	9	87	121	248	457
Germany	168	111	27	301	5011	9013	3454	17478
Greece	3	16	1	19	14	320	33	367
France	249	283	183	678	6454	10392	9758	26604
Ireland	3	33	2	37	30	732	67	829
Italy	18	163	8	187	149	3633	337	4118
Luxemburg	3	1	3	7	63	80	195	338
Netherlands	25	32	17	71	631	1306	760	2698
Portugal	0	0	0	0	1	4	1	6
Spain	101	138	45	276	1417	3855	1999	7270
United Kingdom	167	175	64	394	2371	5141	2446	9958
Total	797	984	400	2101	17337	36129	21645	75110

Table B.4: Specific charge of 16 Ecu/ton, industrial waste, with trade

	Volumes (mln ton)					Costs (mln ECU)			
	Trade	Recycling	Landfill	Proces.	Total	Recycling	Landfill	Proces.	Total
Belgium	-15	47	27	42	122	890	1385	2113	4388
Denmark	1	3	3	5	9	71	191	277	539
Germany	-65	136	83	21	301	3758	6760	2591	13109
Greece	75	3	84	9	19	14	1717	413	2144
France	-130	199	223	158	678	5157	8188	8409	21754
Ireland	59	3	85	10	37	29	1886	446	2362
Italy	54	17	210	16	187	144	4691	707	5542
Luxemburg	2	3	3	3	7	48	208	221	477
Netherlands	2	21	32	25	71	525	1297	1116	2938
Portugal	16	0	15	1	0	1	478	73	552
Spain	3	94	143	52	276	1264	3980	2314	7558
United Kingdom	-1	157	171	81	394	2123	5025	3114	10262
Total	211.3 (315.0)	683	1079	424	2101	14022	35807	21794	71624

Table B.5: Cost-equalizing charge, municipal waste, no trade

	Volumes (mln ton)				Costs (mln ECU)			
	Recycling	Landfill	Processing	Total	Recycling	Landfill	Processing	Total
Belgium	2	1	1	4	40	61	72	173
Denmark	2	1	1	3	33	33	75	142
Germany	18	14	3	34	430	893	423	1746
Greece	2	1	1	4	37	69	44	150
France	11	5	8	23	266	301	434	1000
Ireland	1	0	0	2	18	28	17	63
Italy	12	7	6	24	285	447	284	1015
Luxemburg	0	0	0	0	2	2	6	10
Netherlands	5	2	3	9	107	127	136	370
Portugal	1	1	1	3	27	57	52	136
Spain	7	4	5	15	126	251	230	607
United Kingdom	11	6	9	24	138	404	327	869
Total	73	41	40	146	1508	2672	2101	6281

Table B.6: Cost-equalizing charge, municipal waste, with trade

	Volumes (mln ton)					Costs (mln ECU)			
	Trade	Recycling	Landfill	Proces.	Total	Recycling	Landfill	Proces.	Total
Belgium	0.1	2	1	2	4	38	63	80	181
Denmark	-0.2	1	1	1	3	30	33	73	136
Germany	-1.7	17	13	4	34	406	816	476	1698
Greece	0.4	2	1	1	4	37	75	63	175
France	-1.3	11	5	8	23	248	293	402	944
Ireland	0.4	1	1	1	2	18	33	37	88
Italy	0.4	12	7	7	24	284	452	303	1039
Luxemburg	0.1	0	0	0	0	1	3	13	18
Netherlands	0.3	5	2	3	9	105	132	153	391
Portugal	0.1	1	1	1	3	27	58	58	143
Spain	0.3	7	4	6	15	124	256	247	627
United Kingdom	1.0	11	6	9	24	138	416	364	919
Total	3.1 (4.0)	71	40	43	146	1457	2632	2270	6359

Table B.7: Cost-equalizing charge, industrial waste, no trade

	Volumes (mln ton)				Costs (mln ECU)			
	Recycling	Landfill	Processing	Total	Recycling	Landfill	Processing	Total
Belgium	62	28	40	122	1292	1848	2004	5144
Denmark	4	2	4	9	87	112	248	447
Germany	139	132	37	301	3075	8620	4664	16359
Greece	9	6	5	19	179	393	242	814
France	251	180	309	678	6720	11707	16477	34904
Ireland	15	13	12	37	366	828	524	1718
Italy	76	63	59	187	1807	4112	2618	8537
Luxemburg	3	1	4	7	59	69	227	355
Netherlands	27	20	30	71	705	1273	1330	3307
Portugal	0	0	0	0	2	4	4	9
Spain	130	70	94	276	2368	4558	4184	11110
United Kingdom	190	97	133	394	3198	6292	5137	14627
Total	907	611	728	2101	19857	39816	37658	97331

Table B.8: Cost-equalizing charge, industrial waste, with trade

	Volumes (mln ton)					Costs (mln ECU)			
	Trade	Recycling	Landfill	Proces.	Total	Recycling	Landfill	Proces.	Total
Belgium	-1	59	29	41	122	1220	1864	2068	5152
Denmark	0	3	2	5	9	77	121	288	487
Germany	-15	133	138	18	301	2823	8997	2296	14117
Greece	9	9	8	14	19	179	504	623	1306
France	-42	236	174	283	678	6309	11362	15065	32736
Ireland	9	15	14	20	37	365	939	903	2207
Italy	8	76	65	67	187	1803	4221	2988	9012
Luxemburg	-0	3	1	4	7	48	74	250	373
Netherlands	8	27	21	38	71	697	1377	1687	3761
Portugal	1	0	0	1	0	2	21	76	98
Spain	6	129	71	101	276	2345	4651	4502	11498
United Kingdom	17	190	100	151	394	3198	6517	5804	15520
Total	58.2 (68.9)	881	624	744	2101	19067	40650	36550	96267

6 Pricing Instruments for Transport Policy

Bert van Wee

Introduction[1]

Traffic is very important for society: it enables people to engage in activities like living, working, visiting other people, and going to concerts at different places. In developed economies, on average people travel one hour per day (Zahavi, 1979). Transport is also needed to bring goods from one place to another. Without motorized transport our society would be completely different. The importance of transport is also reflected in its contribution to the Gross Domestic Product (GDP). In the EU the direct contribution is seven percent with further, far-reaching, secondary links through personal consumption of transport service (Button, 1993).

The traffic and transport sector[2] causes many environmental problems (see OECD, 1988, for a general view). The emission of CO_2 possibly results in climate change, the emission of NO_x and SO_2 causes acidification, VOC contributes to high ozone concentrations on ground level and reduces crop growth and many people are bothered by traffic noise. Emissions from motor vehicles (e.g. CO, fine particulates, NO_x) cause or contribute to a wide range of health problems (see Walsh, 1990, for an overview). Whitelegg *et al.* (1993) conclude that living along heavily trafficked streets significantly increases illness. The transport system contributes to the depletion of fossil fuels and of raw materials by using them for vehicles and the infrastructure. In the EU every year 50,000 people are killed and more than 700,000 injured in road accidents (statistics for 1990) (Kageson, 1993; based on ECMT). Finally, the transport system needs land resources, e.g. for roads, rail lines, harbors and airports, both directly for the infrastructure itself, and indirectly, because it limits land use in the area around the infrastructure. This indirect land use is much higher than the direct land use. Land use itself is not an environmental problem, but if nature areas are transformed to infrastructure, it is. And if the landscape is cut into many pieces by the infrastructure, resulting in smaller habitats for animals and in visual pollution, it is. Because of the many environmental problems related to traffic and transport and its relatively high share in many emissions and problems, the sector plays an important role in both policy making and environmental science.

In recent years the EU has changed its policy on the environment, including the environmental impact of transport: it is trying to use pricing instruments much more than before, to reach the targets and to internalize the external costs like the costs to the environment (see below). This chapter discusses the alternatives and problems of the EU in formulating a transport policy in which environmental problems are internalized by fiscal instruments: if the EU wants to make more use of pricing instruments, what are the alternatives? Instruments are regarded as fiscal if they directly change fixed and/or variable costs of vehicle use and ownership. This chapter focuses on the environmental and - to a lesser extent - congestion policy. The focus on environmental issues is evident. Attention is paid to congestion because fiscal instruments seem to be (potentially) very effective in reducing congestion and because of its environmental impact. Besides, this chapter is limited to road transport since this sector plays a dominant role in the environmental problems of intra-EU transport and because EU policy focuses mainly on road transport.

In the next section we show that the general goal of transport policy is to find the right balance between the costs (including external costs such as environmental costs) and the benefits. Several instruments - including pricing instruments - are available to find this balance. The third section shows that environmental costs need to play a prominent role in transport policy because traffic contributes to several environmental problems and has a high share in several emissions. In addition, an increase in the transport of both people and goods is expected. Next, the EU policy relevant for transport and the environment is described. It appears that until now, most progress has been made with respect to emission regulations. At the same time, however, extra (international) transport is generated as a result of the 'open border' policy. Since 1992 official EU policy has been to internalize external costs, such as the costs to the environment. Pricing instruments have to be used more than in the past. The fifth section discusses several pricing instruments, mainly with regard to variable costs of vehicle use and to a lesser extent with regard to fixed costs and subsidies. The sixth section shows that pricing instruments are not always the best solution for all problems; sometimes regulation is preferred. Finally, the main conclusions are presented.

Goals for Traffic and Transport

Goals

Before discussing policies and instruments we need to know what goals we want to achieve. In this section these goals are discussed briefly, in order to define what an optimal transport system is. Once the goals are specified we can develop a strategy and policy to reach these goals. This means that concrete objectives and instruments have to be chosen. A general framework linking objectives and instruments will be presented.

The general goal for a transport system is (a) to allow people to optimize the fulfillment of their location-related needs and (b) to optimize the transport of goods between different locations. From an economic point of view the transport system should be improved (changed) until an optimal point is reached. This point is reached if marginal social benefits equal marginal social costs. In other words, there is a benefit component and a cost component.

As far as passenger transport is concerned, the benefits of a transport system are the result of providing people with opportunities to fulfill their location-related needs. Human beings have many needs, wants and desires, resulting in many activities. Most activities, such as working, living, shopping and recreation, are location related. The traffic and transport system makes it possible to travel between these locations. Notice that some activities are location related in the current situation, but may be not-location related in the future. For example, telematics could reduce the need to travel.

This general goal supposes that people only travel because they want to be in different places. Apart from this intermediate goal, two other motives can be mentioned. First, sometimes people travel only because they want to travel. Travelling then is a goal in itself, a kind of recreation. Second, some psychologists emphasize the psychological functions of cars (or bikes, motorbikes). They refer to needs such as status, territorial needs and power.

With regard to the transport of goods, the benefits of a transport system result from the opportunities to transport them between different locations. These can be locations where raw materials are produced, where intermediate products and final goods are produced, distribution locations, shops, locations of users (firms, families) and locations where waste products are collected, recycled, incinerated or land filled.

Almost all forms of transportation cost money. Reducing transport costs as much as possible is therefore an important goal, both for users as well as for policy makers. Total costs include different components. First, the *direct costs* related to building the infrastructure, producing vehicles and operating the system. These direct costs can be divided into fixed and variable costs. Costs are variable if they depend on the level of use. Costs are fixed if they are not influenced by the level of use.

Besides direct costs, there are other *in-system costs.* Travelling is risky, mainly because of (differences in) speed, inherent conflicts and imperfect protection of vehicles against accidents. Damage to goods, due to accidents also incurs costs. Traffic therefore generates accident costs. Furthermore, transport takes time, and time is money, both for people and for goods. This presents us with two options for reducing the in-system costs: improving safety and reducing travel times (e.g. by reducing congestion).

Third, there are *costs outside the transport system:* environmental costs, such as noise nuisance, local air pollution, acidification and climate change. This brings us to another goal of the transport system: reducing environmental impact. Part of the in-system costs and all costs outside the transport system (environmental costs) are called external costs or externalities: effects on a receptor that are not taken into account in decision making by the actor.

Balancing the Costs and Benefits

The purpose of transport policy is to find the right balance between benefits and costs: the optimal situation has to be found. But what is optimal? In general terms 'optimal' means that marginal social benefits meet marginal social costs, including external costs. So, building more road infrastructure improves the transport system only if marginal benefits (e.g. in reduced travel time and costs) are higher than the marginal costs (such as fixed and variable costs of the infrastructure, costs of increased emissions and noise). In other words, reducing travel time 'at any price' is not optimal. Transport economists have shown that there is an optimum for reducing congestion by building an infrastructure and, at the same time, admitting congestion. At this optimum there is still a certain level of congestion. Costs of building more than the optimum amount of infrastructures are higher than the savings because of reduced travel time.

To find the optimal situation, all relevant aspects have to be considered, including environmental impact: a cost-benefit analysis has to be made. This presents two main problems. First, there is the problem of measuring costs: how to price the 'unpriceable'? There is considerable consensus on some valuation methods, e.g. the pricing of travel time, although some have argued that this method is also tricky. It is beyond doubt that the monetary valuation of deaths, injuries, emissions of pollutants like NO_x, SO_2 or CO_2 is much more difficult. Although economists have developed several methods, pricing accidents and the environment is, first of all, a political choice. Some people have objections to putting a price on the environment. The best way to deal with this objection in cost-benefit analysis may be to define the maximum acceptable overall emission levels (not only for transport). Then the costs of limiting emissions to this maximum level have to be estimated. This estimate can be used in the cost-benefit analysis.

The other problem is the measurement of the benefits of a transport system. Some people argue that for passenger transport the total distance travelled is a good indicator. Others argue that the amount of money spent on mobility is a good indicator. This is questionable. If, for example, the number of hospitals is reduced, on average people have to travel longer to visit their family and friends if they are in the hospital. Nevertheless, they would prefer shorter trips. In future, scientists may develop indicators for the level of mobility needs that are satisfied by a transport system. Perhaps the number of trips and an additional benefit for the consumer surplus is a good indicator. These indicators can then be related to the costs of the system, including external costs. Optimizing the indicator and relating it to other (non-transport) indicators for the needs of people and the related costs may be a more satisfactory way to judge a transport system than simply developing a system that reduces emissions by a certain percentage.

Objectives and instruments

Once the goals are specified, policy objectives and instruments necessary to achieve them have to be chosen. In this section an *objective* refers to *what* is wanted (e.g. a shift from car use to public transport, the use of cleaner technology); an *instrument* refers to *how* this is to be achieved (e.g. changes in prices, regulations).

Some of the goals will be related to the environment and to congestion. The environmental impact of traffic and transport and the congestion level can be influenced by (realization of) several kinds of objectives, related to the total volumes (e.g. less mobility or less car use), the modal split (e.g. a shift from car to public transport), the techniques used (e.g. the use of three-way catalytic converters), the efficiency of vehicle use (e.g. fuller trucks, higher occupancy rates of cars) and the way vehicles are used (behavior, e.g. the driving speed of cars). There are also several kinds of policy instruments available. The most important instruments are infrastructure policy, information and organization, marketing, land use planning, pricing instruments and restrictions/regulations. Table 6.1 shows how these instruments can contribute to the objectives, thereby changing the environmental impact of traffic and transport.

Table 6.1: Relationship between instruments and objectives.

	volume	modal split	technology	efficiency	behavior
infrastructure	*	*	*	*	*
information and organization		*		*	
marketing		*			
land use planning	*	*		*	*
pricing	*	*	*	*	*
restrictions	*	*	*	*	*

Source: Blok and Van Wee, 1994.

Table 6.1 shows that pricing instruments are only one of several possible instruments, but they are important because they can be used for all objectives. Higher prices can reduce car use (volume), result in a shift toward public transport (modal shift), influence driving behavior and improve the efficiency of the use of trucks. Pricing policy can speed up technological improvements.

Once the goals of a transport system are defined and the choice is made that external costs have to be internalized by means of pricing instruments,[3] the question is which prices have to be increased or decreased by fiscal instruments. Many external effects are related to vehicle *use*, more than to vehicle *ownership*. Therefore increasing variable costs seems appropriate for reaching the goals. Maybe fixed costs can be decreased. Whether or not this is advisable according to economic theory depends on how to price external costs of car ownership.

Traffic and the European Environment

The traffic and transport sector has a significant share in emissions of several pollutants. Table 6.2 shows total transport emissions (road traffic and non-road traffic) as a percentage of total emissions for some (groups of) countries and gives noise nuisance data. In the OECD, transport (including marine bunker fuel use) is now responsible for more than one-third of final energy use. It is the largest final energy use sector and its share is growing (OECD, 1993a).

Table 6.2: Some environmental impacts of transport in industrialised countries.

	North America	OECD Europe	Japan	OECD
Emissions[1)]				
NO_x	47	51	39	48
CO	71	81	n.a.	75
SO_2	4	3	9	3
Particulates	14	8	n.a.	13
HC	39	45	n.a.	40
Noise[2)]	19	53	36	110

[1)] Transport emissions as % of total emissions.
[2)] Population exposed to road traffic noise over 65 dB(A), in millions.
Source: Button, 1993; recent OECD data.

In the EU, total transport of goods (in tonkilometers) increased by more than 50 percent between 1970 and 1990. Passenger transport increased by 85 percent in this period. The share of the environmentally unfriendly modes (road, air) increased sharply, while the share of rail and water decreased. For example, road haulage more than doubled. To date, energy consumption of the transport sector represents 30 percent of total final energy consumption in the EU. Road transport consumes over 80 percent of the total final energy used in the transport sector and contributes over 75 percent to its total CO_2 output

(CEC, 1992b; For trends in other world regions, see MacKenzie and Walsh, 1990).

Between 1980 and 1990 SO_2 emissions from mobile sources in West Germany, France and the United Kingdom decreased by eight percent. In the same period NO_x emissions from mobile sources in these countries increased by 25 percent (calculations based on statistics of Umweltbundesamt, CITEPA, National Atmospheric Emissions Inventory).

In the future both car use and the use of trucks are expected to increase significantly. If incomes rise, car ownership is expected to grow even more: the income elasticity for car ownership is more than one (Van Gent and Rietveld, 1990), but will decrease as incomes approach those related to the saturation level for car ownership. In its *Environmental Outlook for Europe* for the period 1990-2010, the RIVM (1992) forecasts a growth in car use in Western Europe of 55% and a growth in total freight transport of 40% (tonkilometers). This trend will contribute to rising emissions and energy use. On the other hand, higher emission standards and technological improvements may be expected, resulting in cleaner technology and lower emission factors (emissions per kilometer). Besides, new cars and other vehicles will be more fuel efficient than present vehicles.

In the CEC scenario study on energy in Europe, future levels of emissions are forecast. In the 'conventional wisdom' scenario, SO_2 emissions of transport in EU-12 will grow by 16% and CO_2 emissions by 13% between 1990 and 2010, while NO_x emissions will be reduced by 29%, mainly due to emission standards (CEC, 1990). If expected reductions of sulphur in fuels are realized, SO_2 emissions will be reduced by more than 50% (Van Wee *et al.*, 1993).[4] Theoretically the potential for improving the fuel efficiency of new cars is tremendous (see Lovins *et al.*, 1993). Consequently, emission levels of most pollutants are expected to decrease, but - at least in the Netherlands - not enough to meet most targets. Contrary to most emissions, the CO_2 emission of road traffic in the Netherlands is expected to *increase* by about 20 percent between 1990 and 2010, whereas the target is a *reduction* of about 20 percent[5] (RIVM, 1994). Trends in other EU countries will not differ dramatically from those in the Netherlands.

Congestion has been a common phenomenon in cities and towns for many years. In the last ten years congestion on the motorways has grown rapidly in most western countries, implying a loss of welfare. In the Netherlands between 1986 and 1990 the costs of congestion as a result of extra travel time increased from 775 to 1060 million guilders (NEA, 1991). Additional congestion costs can also be caused by extra emissions (environmental impact) and increased costs of accidents. The expected growth in road traffic will cause more and more congestion unless policy changes the situation.

The relationship between congestion and the environment is complex. Congestion results in higher emission per kilometer of pollutants, like NO_x and VOC, and in higher energy use per kilometer. Reducing congestion - ceteris paribus - reduces the environmental impact of transport. The problem is the ceteris paribus assumption: it does not exist. Reducing congestion reduces travel times and so generates more traffic. The balance between positive and negative environmental aspects of reducing congestion depends on the specific local or regional situation. Simply not solving congestion because of a possible positive environmental balance is not likely to result in optimal welfare. It seems a better policy to reduce congestion to a certain (optimal) level, while at the same time reducing environmental impacts using additional instruments.

External Effects of Transport

In recent years several studies have been carried out to evaluate the external effects of the traffic and transport sector in monetary terms (external costs) (see Kageson, 1993; Verhoef, 1993; MacKenzie *et al.*, 1992; Boneschansker *et al.*, 1994; Rothengatter, 1993; Quinet, 1990). The studies conclude that traffic does not pay for itself, including the costs of traffic accidents, emissions, noise nuisance and congestion. The level of external costs depends to a great extent on the methods used to measure them.[6] Kageson calculated the external costs of transport for EU countries. He concludes that in 1992 taxes on gasoline and diesel in the EU countries were between 25 and 60 percent of the taxes needed for internalizing the present external costs. In contrast to other studies, Kageson valuates CO_2 emissions. He also concludes that the external costs differ considerably among countries, mainly because of differences in costs of accidents.

The fact that the transport system pays for only part of its external costs can be changed, both by regulation and pricing instruments. Regulation can reduce external costs: emission regulation can reduce emission levels and therefore environmental costs. Pricing instruments can have two types of effects. First, as a result of higher prices, specific external effects as well as the transport volumes can decrease, which reduces environmental costs. Second, if pricing instruments are used, a price (tax) has to be paid for the remaining external effects. Then the extra tax proceeds can be used to repair environmental damage, to pay the victims or to lower other taxes, e.g. income taxes. The level of pricing instruments can be related to the level of external costs, or alternatively to the desired *effect* of pricing: prices are set as high as needed to achieve the desired effects. A scientific reason for not basing the price levels on external costs is the fuzziness of the calculations. Besides, experience shows that in practice the opportunities to internalize externalities by pricing *or* regulation in the transport sector are limited. Some historical taboos exist in Member States of the European Union that yield a high resistance of the industry and of consumers to any kind of limitation of their private wants in transport (Rothengatter, 1993).

The Traffic and Transport Policy of the European Union

The EU Policy up to 1992 and Its Results

The European Union is aware of the environmental problems of transport, as described before, and (especially since 1992) of the fact that environmental problems are external problems that can be internalized. This section focuses on concrete policies of the EU, as far as relevant for the environmental impact of transport: the general environmental EU policy, the EU transport policy and other EU policies. First, EU policies up to 1992 will be described, as well as the results of these policies.

The EU policy on the environment and sustainable development is described in so-called action programmes. Although they do not describe transport policy in detail, they will be discussed briefly. In 1972 environmental policy first got an official position in EU policy, leading to the First Action Programme that was published in 1973. The first two action programmes (1973-1977; 1977-1981) focus on (among other things) prevention of pollution, the 'polluter pays principle' and international environmental cooperation. In the Third Programme (1982-1986) the wish to integrate environmental policy in other policy sectors was launched. In addition, the programme emphasizes the interaction between social and economic developments and the environment. The Third Programme makes numerous references to transport issues, covering pollution related to both vehicles and infrastructure, and proposes the development of new, low-polluting technologies and the maintenance of an inventory of the best available low-pollution technologies. The Fourth Programme (1987-1992) focuses on prevention and reduction of pollution of air, water and soil by source-oriented measures. The programme emphasizes the importance of high environmental standards on the EU level and its emphasis on the transport sector is similar to that of the Third Programme.

Until a few years ago the implementation of EU policy on transport and the environment focused on regulation, mainly with regard to emission standards and fuels (lead in gasoline, sulphur). Standards have been set to reduce emissions of traditional types of atmospheric pollution, i.e. smoke, sulphur dioxide and floating particulates (Vougias, 1992). The first motor vehicle emission standards date from 1970. The first EU noise standards for motor vehicles also date from 1970. All countries have the same standards. To meet the 1993 standards[7] gasoline cars need a three-way catalytic converter which reduces emissions of NO_x, VOC and CO by approximately 80 percent. In 1978 the Council adopted a directive on the maximum authorized lead content in regular or premium gasoline. In 1985 and 1989 further directives were launched (lead, sulphur) (Rotterdam Transport Centre and Erasmus Centre for Environmental Studies, 1991). Also progress has been made in the harmonization of motor fuel taxes in EU Member States (Smith, 1994). In the first quarter of 1993 gasoline taxes in EU Member States ranged between 66 and 78 percent of total gasoline prices, while in most other OECD countries this percentage was much lower and variable, with the lowest percentage in the USA (30%) (OECD/IEA, 1993).

Other relevant EU policies include the *urban policy*, which, according to the urban 'Green Book', focuses on comprehensive policies to solve urban environmental problems related to transport. The role of the EU, however, is restricted to encouraging local authorities to integrate land use, public transport and road planning, by providing information and contributing to the cost of pilot projects. Therefore, the urban policy of the EU has had only limited effects. For international *road haulage* a quota system of EU licenses was first introduced in 1969. Since then, the number of licenses has increased gradually, which can be considered a step towards the (current) policy of liberalization (see below).

The general 'open border' policy has been very important for transport. This policy (the Europe 1992 program, or the completion of the Internal Market) was developed for economic reasons: the economies of individual Member States would benefit if fiscal and other barriers at borders were eliminated, resulting in the free movement of people and goods. The open border policy has been very successful and has very likely been good for the economies of Member States. One result of the policy has been a greater increase in transboundary transport than would have been the case without the open border policy. This extra transport results in extra emissions. Related to the open border policy, the harmonization of legislation and the increase of EU influence on national policies, results in certain restraints on individual Member State policies. This change has both positive and negative environmental impacts. The environmental damage in countries with traditionally limited environmental policy for transport is decreased by this change, while the improvements for more progressive countries are limited or may even be negative.

Current and Future EU Policy

In 1992 two reports on transport were published: the 'Green Paper' and the 'White Paper' (CEC 1992b), as well as the Fifth Action Programme on the Environment (CEC, 1992a). In these policy plans the general EU strategy for *sustainable mobility* is described. Many aspects of the proposed policy had already been addressed in the report of the Group Transport 2000 Plus (1990). The EU admits that until 1992 "the progress in the realisation of the Community's common transport policy was slow". Present policy is to change this: the EU wants sustainable transport. Some key elements of the present policy are the creation of a more open market, free from unnecessary restrictions, fair competition, improvements in competitiveness, transport systems, safety and reliability, and less environmental damage.

The EU states that "not only will transport need to respond efficiently to market demand, it will need to do so at the lowest possible cost for society, taking fully into account environmental costs. This requires the pursuit of efficient policies in the fields of pricing, including the internalisation of external costs". This policy of internalizing the external costs is the current policy for other sectors as well (see the Fifth Action Programme). Until now fiscal instruments have been used mainly to generate funds for governments; internalizing external costs and regulation of traffic demand and emission via pricing have received very little consideration. So, recent EU policy on pricing means a significant change in the motives for pricing policy. The EU states that fiscal and economic instruments have several advantages. They can be focused on the source of pollution and give equal treatment to different sources and uses, they can be made equivalent in amount to the external costs associated with pollution and are thus a particularly direct application of the principle that the polluter should pay through the internalization of external costs, and they are flexible in the sense that they can be used as an incentive to avoid or limit certain behaviors. Tax incentives can directly encourage the user, for example, to opt for a more energy-efficient vehicle, while a progressively heavier charge on the use of less energy-efficient vehicles can discourage their use.

In the Fifth Action Programme on the Environment the EU sets some targets for total emissions in the future (see Table 6.3). Specific quantitative targets for the transport sector are not given. If transport has to contribute to the reductions to the same extent as other sectors, the targets for NO_x, SO_2 and VOC may be within reach. For example, stricter emission standards for trucks in the year 2000 may reduce the emissions of NO_x and particulates per kilometer by more than 50 percent. Recent evaluations show that - if present policies are implemented successfully, including the expected EU vehicle emission standards - the emission targets for NO_x, SO_2, and VOC will be reached (RIVM, 1994). The CO_2 target, however, will be very difficult to meet. For a stabilization of CO_2 emissions in the transport sector the EU needs an adequate policy. Very likely both the growth in vehicle use has to be reduced and the fuel efficiency (energy per vehicle and/or tonkilometer) has to be increased significantly. For such a policy, pricing instruments can be very helpful (see below). In future, the EU also wants to introduce standards for energy consumption. Such standards might be effective as well.

Table 6.3: Some EU targets for total emissions.

CO_2	stabilization in 2000 at 1990 levels (+ progressive reductions by 2005 and 2010
NO_x	reduction of 30% between 1990 and 2000
SO_2	reduction of 35% between 1985 and 2000
VOC	reduction of 30% between 1990 and 2000

The EU infrastructure policy is also relevant for transport. The EU policy on infrastructure is described in the White Paper on *Growth, Competitiveness and Employment* (CEC, 1993). The EU wants to develop trans-European infrastructure networks, because it wants better, safer and cheaper transport, for land use planning reasons and because it wants to improve relations with Eastern Europe. The EU wants to stimulate the construction of these networks by removing legislative obstacles, encouraging private investments in infrastructure projects, and selecting projects from plans. Competition, employment and economic motives play an important role in this White Paper, while the environment is hardly mentioned. From an economic point of view, removing legislative obstacles is a good action, but this will also encourage international transport and - ceteris paribus - result in higher emission levels. The construction of new infrastructures will also encourage transport. Investments in railways and waterways may reduce emissions of road traffic. Investments in road transport will result in higher emissions. Therefore from an environmental point of view, investments should be limited to only those projects where the marginal economic benefits more than compensate for the extra environmental costs. Unless all external costs are internalized, other instruments to meet the environmental standards should be introduced or strengthened.

The EU plays an active role in infrastructure policy because of its structural funds for investments in infrastructure (roads, railways, waterways). For the period 1994-1999, the funds for financing trans-European networks are 5.3 billion ECUs per year. The EU can use road pricing for the networks that will be (partly) financed by the EU.

The Role of the EU and of (sub)National Authorities

There are several ways in which the EU influences the individual Member States. First of all, individual Member States are obliged to implement EU regulations, e.g. on emission standards. This means that countries with traditionally low emission standards will have to implement higher standards. More important is that countries that want higher standards are not allowed to set them. This seems important because the EU has implemented stricter standards much later than technically possible, and later than in e.g. the USA. Individual Member States can only encourage people and companies to buy vehicles that already meet future EU standards.

Next, the EU can stimulate individual Member States to develop and implement certain policies. For instance, the EU wants to promote collective and environmentally friendly means of transportation and states that the national and local authorities will play an important role in the use of the related instruments (e.g. subsidies). The EU can also provide the framework for the use of tax incentives which make environmentally friendly transport solutions more attractive. The EU mentions, for example, congestion pricing as a useful instrument to reduce congestion. The EU's role is to provide common framework and guidelines for the use of economic instruments.

Third, the EU plays an active role in fiscal policy, e.g. taxes on motor fuels and VAT. Table 6.4 shows taxes on gasoline (regular) and diesel fuel in four major EU Member States in 1973, 1983 and 1993. Obviously, the differences in taxes on gasoline and diesel decreased considerably between 1973 and 1993, partly due to EU harmonization. In contrast, differences in rates of VAT (or general consumption taxes) still vary considerably between energy products. In six Member States the VAT rate on diesel fuel is 0%, while in the other Member States diesel is taxed according to the general VAT rate. Gasoline is rated in most countries according to the general rate (exceptions: Belgium, Ireland and Luxembourg).

From an environmental point of view, there is hardly any reason for vast differences in the rates of VAT and other taxes between energy products and between Member States. Although considerable progress has been made in the EU in harmonizing the taxes on motor fuels, there is still a long way to go. The most recent proposal explicitly adopted environmental considerations in setting target levels for the excise rates on several mineral oils, such as gasoline, diesel, liquid petroleum gas (LPG), and light and heavy oils throughout the EU. The response called for is a tax rate reform: the currently existing tax rates on some energy products have to be reformed by individual Member States in order to bring them in line with the proposed levels of taxation for the Union as a whole. The ultimate goal is the harmonization of excise duties between Member States, reducing

Table 6.4: Taxes on petrol (regular) and diesel in four EU Member States, 1973, 1983 and 1993 (in US $ per gallon).

	Regular petrol			*Diesel fuel*		
	1973	1983	1993	1973	1983	1993
France	0.38	1.01	2.86	0.23	0.65	1.22
Italy	0.29	1.61	3.06	0.13	0.33	1.68
UK	0.26	1.08	2.26	0.26	1.05	1.46
Germany	0.70	0.92	2.62	0.66	0.82	1.29
max/min ratio	2.69	1.75	1.35	5.08	3.18	1.38

Source: Smith, 1994.

unequal competition between Member States. The environmental objective provides an additional consideration (Vollebergh, 1994).

The harmonization of fuel taxes has resulted in higher tax rates in countries that traditionally had low taxes. On the other hand, harmonization imposes a limitation on Member States that want higher taxes than the maximum. The same relation between the EU and Member States is expected for future harmonization of taxes on vehicles, and for fiscal incentives which may be used in the future to encourage the use of alternative and cleaner fuels. Such incentives will very likely be harmonized by the EU. On the other hand, land use planning, spatial planning and parking policy are, to date, primarily a national and/or local matter and will remain so in the future.[8]

In general the conclusion is that the EU can improve environmental impacts (reduce negative external effects) by EU policies and by stimulating individual Member States to implement national, regional or local measures. On the other hand, the options for individual Member State policies are limited by the EU (e.g. higher standards and fuel taxes above a certain maximum). A change in EU policy might reduce these limitations (see below).

Prospects for Pricing Instruments in the European Union

Improving European policy

Although much progress in EU transport policy has been made since 1992, further improvements are possible. First of all, quantitative targets for transport, not only for 2000 but also for 2010 and maybe further into the future, can be set. This is especially advisable for transport, since infrastructure planning and the development and penetration of new vehicles take many years. Translating the current policy into more concrete actions will also increase the chances for success. Finally, internalization of external effects - including CO_2 emissions - is needed, as well as a concrete policy aimed at enhancing the role of pricing instruments. This section discusses how the EU can best further the use of pricing instruments.

Generally speaking, pricing instruments may influence the level of (specific) *variable costs,* i.e. those direct costs which depend on the level of use (e.g. traditional fuel taxes or innovative instruments like road pricing), or the level of (specific) *fixed costs,* i.e. those direct costs which are not influenced by the level of use (like taxes on new vehicles or annual recurrent taxes on vehicles). Other instruments may influence the level of fixed and variable costs at the same time. Finally, we distinguish subsidies as a special category. In this chapter we evaluate pricing instruments almost exclusively on their effectiveness in reducing external effects like environmental damage and congestion. It is important to note that pricing instruments may also contribute to other goals, like generating funds for the government, financing of infrastructure and land use goals.

With regard to pricing instruments, the EU and individual Member States have to play different roles. Traditionally, the EU policy was mainly focused on harmonization of existing taxes, e.g. fuel taxes, taxes on new cars and other vehicles, and recurrent taxes on vehicles. The highest priority is to reorient this harmonization policy to the goal of internalizing external costs. A choice has to be made about which external costs should be internalized by fiscal instruments related to fixed costs (taxes on new cars, recurrent taxes) and related to variable costs (fuel taxes). Then minimum levels could be set, corresponding to a 'consensus level' of external costs that appears in most (if not all) EU Member States. Sometimes it is argued that the EU also has to set maximum levels, because of harmonization. However, there seems no need for maximum levels, as individual Member States can define their own level of taxes, taking into account border effects if neighboring countries have lower taxes, their own views on external costs and how to price them and their overall financial position. An additional role for the EU with regard to existing taxes, is to provide information on external costs and how to price them, in order to help Members States to set proper tax levels. Finally, the EU can act as independent mediator in case of tax disputes between countries.

Besides reorientation of existing taxes, several new pricing instruments may contribute to the internalization of external effects, like road pricing, parking fees, tax differentiation or subsidies. The main role of the EU with regard to such instruments can be to stimulate individual Member States in the first place. Knowledge of the effects of these instruments may be limited in some Member States, so the EU can play a role in providing information and guidelines on where and how to introduce them, and in financing studies on their effects. If new roads are (partly) financed by the EU, they can also play an active role in introducing road pricing, at least on the roads financed by the EU, but perhaps also on 'interacting' road sections. And finally, the EU can increase the opportunities for Member States to use certain pricing instruments. To date, individual Member States are allowed to use tax differentiation between newly sold cars, only to stimulate the selling of cars that already meet the subsequent EU standards. It is questionable whether there is any need for this limitation: individual Member States should be able to stimulate any technology that performs better than current standards. The only problem may be the possible abuse of tax differentiation to strengthen the market position of national car manufacturers if they have some technology that is not available in other countries. One way to handle this may be that tax differentiations will need to be announced at least some years in advance, so that car manufacturers can be prepared.

Pricing Instruments in Detail

Now that some general suggestions for EU pricing policy have been discussed, several pricing instruments will be presented in more detail. The headings of the instruments refer to the unit that is priced. That is, the section 'fuel taxes' only focuses on the impact of taxing fuels, 'emission charges' only focuses on instruments that charge according to emission levels. However, fuel taxes can be regarded as an indirect way of charging emissions. This example shows that several instruments can be used to obtain a desired effect.

Apart from the instruments discussed below, other pricing instruments are proposed in the literature, e.g. tax incentives to encourage carpooling and to encourage people to reduce their home-work distance (by moving closer to their work or by choosing a new job closer to home), or permit systems to reduce rush-hour traffic. These instruments will not be discussed in this chapter.

Fuel taxes - Raising levies on motor fuels is probably the most discussed type of pricing instrument, both in policy and economics. The main advantage of higher levies on motor fuels is that it reduces both car use and the energy use per kilometer: the effect of higher fuel prices on energy efficiency is greater than the effect on car use (kilometers driven) (Goodwin, 1992; NEI, 1991). Kageson (1993) mentions two review studies. Combining the two studies, it can be concluded that long term elasticities of energy use for gasoline prices range between -0.65 and -1.0. One of these studies (NEI, 1991) concludes that long term elasticities of car kilometers range between -0.1 and -0.3. So, an increase in fuel prices of 1 percent results in a reduction in car kilometers of 0.1 to 0.3 percent and of energy use of 0.65 to 1.0 percent. Higher levies on fuels can also reduce road haulage. But the elasticity is supposed to be even lower than for car use.

Taxing fuels at a much higher rate has been discussed recently in order to reduce CO_2 emissions.[9] The OECD (1993a) concludes that in economic terms the most efficient way to reduce CO_2 emissions would be to tax all fuels, in all sectors, throughout the world, according to their carbon content. Montgomery (1992; cited in Smith, 1994), states that "a mid-range estimate is that a tax of $200 per ton of carbon would be required to hold emissions to 20 percent below their current levels through the first quarter of the next century". Smith (1994) concludes that such a tax would increase gasoline prices in most EU Member States by 50 to 100 percent.

Taxing fuels only on their carbon content has the advantage that there is a direct link to the CO_2 emissions caused by using the fuels. A disadvantage is that this tax is only related to CO_2 and not to other relevant pollutants like NO_x and VOC. Relating fuel taxes to other pollutants is, however, very complicated because most other emissions (except lead and SO_2) depend very much on the vehicle used, more than on the fuels themselves. In the past, differentiation of fuel taxes between leaded and unleaded fuels has proven to be successful (UN, 1990; Button, 1993). In the Netherlands, Switzerland, the UK and Germany this tax differentiation has been combined with the banning of normal leaded gasoline, which has further reduced the demand for leaded gasoline (Button, 1993). This example shows that tax differentiation can be combined with other types of regulation. Some have suggested tax differentiation between diesel, gasoline and LPG for environmental reasons (because of differences in emissions). However, this may be tricky as different fuels tend to have a mix of advantages and disadvantages relative to each other. So, some judgement is inherently needed to decide which attributes are preferred (Bailey, 1994). Nevertheless, once the (political) choice is made, tax differentiation of fuels for this reason may be successful.

The main disadvantage of higher levies on fuels is that there is no differentiation between time and place. This makes higher levies on fuels more attractive for environmental goals (e.g. a reduction in CO_2) than for reducing congestion. In the EU countries levies on fuels are harmonized. This means that there are maximum and minimum levies on fuels. This limits the options of the individual EU members. Another disadvantage is that fuel taxes reduce energy use and CO_2 emissions much more than emissions of other pollutants, because the price elasticity of energy use is much higher than that of vehicle use, and because on average emission factors of NO_x, VOC and CO of fuel efficient and non-efficient cars hardly differ. So, higher fuel taxes may be the first best solution for reducing energy use and CO_2, but not for other pollutants.

Road pricing[10] - The main advantage of road pricing is its possible contribution to reducing congestion. As far back as in the 1920's, economists like Pigou (1920) and Knight (1924) recognized that road pricing offered the first best solution for optimizing congested road traffic flows. Recently, the interest in this instrument has grown in the EU as well as in several Member States. The most attractive type of road pricing seems to be a system of Electronic Road Pricing (ERP). The reason for its superiority is that it can be differentiated according to the various dimensions which determine the actual marginal external costs of each trip, such as the length of the trip, the driving time, the route followed and the vehicle used. The Hong Kong experiment has demonstrated that it is technically possible to successfully operate an ERP scheme nowadays. A less sophisticated alternative for ERP might be a toll system. The main advantage is its less complicated character. It has proven to be successful in many countries such as France, Spain, Italy and Greece. The most common reason for charging toll has been to raise funds to pay the infrastructure costs. Nevertheless, a toll system can reduce car use and so congestion and emissions. Access charging is sometimes seen as a special type of road pricing; others discuss it separately. Access charging in urban areas has been successfully implemented in three Norwegian cities.

It is difficult to evaluate the (potential) environmental impact of road pricing. Theoretically some impact can be expected, because car use becomes more expensive and the price elasticity of car use is below zero. A reduction in car use results in a reduction of emissions. Effects of a possible ERP scheme for the Netherlands have been computed, using a national transport model. The simulated system distinguishes between regions and time of day. The overall increase in variable costs is 20 percent. The calculated overall reduction in car use (kilometers driven) is 6 to 13 percent. The more other cost instruments are used (mainly taxes on fuels), the less the additional effect of ERP. The results suggest an elasticity of -0.3 to -0.65.

The effects of road pricing depend very much on local conditions. Elasticities may differ between countries or regions. If there is a good alternative route, the main effect can be a change in routes. If there is no alternative route, but a good public transport alternative, a modal shift can be the main effect. If there is no alternative route and no public transport, the main effect can be a reduction in trips or a change in time of day. Particularly in this situation, but also in other situations, relocation of families and firms may occur; families may move closer to their work. If workers change jobs, their 'search area' can be changed. If congestion is strongly reduced by the road pricing system, the region may become more attractive. On the other hand, if road pricing results in high costs, firms could leave the region and move to less congested areas. This can result in higher car use levels. In conclusion, relocation of firms and families can have both positive and negative effects on car use.

The main objections usually raised against ERP include: (1) privacy considerations; (2) equity considerations; (3) costs of introduction and operation, possibly exceeding the expected welfare benefits; (4) the possibility of car drivers following escape routes (avoiding toll points and increasing external effects in other areas); (5) difficulties associated with the determination of the optimal road price and (6) perverse incentives faced by regulatory bodies (Button, 1992; Evans, 1992). Some of these objections can also be raised to a toll system or access charging. These systems are easier to implement, but may have higher operational costs, increase travel times because paying takes time and are less flexible in tariff structure.

Emission charges and distance charging - Emission charges do not yet play an important role in the field of air pollution control in general or in transport. Theoretically, emission charges are the first best solution for reducing emissions. It is relatively easy to implement emission charges through differentiation of fuel taxes, if emissions only depend on fuel consumption. Examples are lead and SO_2 emissions (see above). However, emissions of most pollutants usually do not depend on fuel consumption, but on many other factors. To date, emission charges on these pollutants are difficult to implement, because of the very high costs of gathering and controlling information on emissions of non-stationary transport vehicles (UNECE, 1994). Perhaps techniques can be developed to gather the required information. In the future a combination of 'emission inspection' and annual safety checks may be useful.

Distance charging means paying for each kilometer driven. This pricing instrument seems, like the preceding one, difficult to implement in the short term, because it presupposes a reliable measurement of distances, covered by each vehicle. One possible solution might be a registration system of kilometer counters, in combination with the annual safety check. From an economic point of view distance charging is an instrument with many benefits. The main advantage is that emissions of NO_x and VOCs are more or less related to the vehicle distance travelled. Therefore distance charging is more satisfactory for reducing these emissions than fuel taxes or road pricing. The best results are obtained if distance charging is related to emission factors, directly or via vehicle size and type. Then distance charges can be an incentive to retrofit vehicles with lower emission abatement technology. They can also be used to make the users pay for the infrastructure and its maintenance (Bailey, 1994). The main disadvantages are the possible sensitivity of this instrument to corruption and the fact that the introduction of a new instrument often meets an acceptability problem: strong opposition may be expected.

Parking policy - Parking policy can influence car ownership and use in general and in local situations (noise nuisance, local air pollution, liveability) and so environmental costs. The most important elements of this policy are higher parking fees and extending the area (number of parking places) of paid parking. Another element of parking policy can be the reduction of parking places, especially in inner cities. However, this is not a fiscal instrument and will not be discussed here. One of the problems of a parking policy in central areas is that it can reduce parking in the paid area but remove it to the surrounding areas. To relieve the situation in these surrounding areas a parking policy is needed in these areas as well, which might remove the problem to yet another area. Another problem can be that if parking tariffs are raised in the central area, a shift from long-term to short-term parking may result and so the number of car trips to and from the area is increased. This increase may cause other external effects that have to be solved. Model simulations can help determine what policy will yield best results for a specific area or region. A concentration of paid parking in a limited number of places with good access and tariffs that cover both internal and external costs seems to be the first best solution for parking problems and congestion in and near town centers. Another problem with parking policy is that in many city and town centers a high proportion of parking places are private off-street facilities, with less opportunity for intervention.

Purchase tax and recurrent taxes on vehicles - The most important 'fixed costs instruments' are purchase taxes on new vehicles and annual taxes on vehicles. Raising these taxes can be an instrument to internalize external effects of car ownership, such as land use consumption and visual pollution. Increasing fixed costs will lead to a decrease in car ownership and may also contribute to a reduction in car use and external costs of car use. Some support for the effectiveness of this policy is the relatively low level of car ownership in Denmark where taxes on new cars are much higher than the EU average. Since car ownership is relatively price inelastic, drastic changes in car costs are needed to have a significant effect.

By differentiation of purchase taxes (or recurrent taxes) on vehicles, people can be encouraged to buy cleaner and more fuel-efficient cars. Such tax differentiations are easy to implement and the additional administrative costs will be minor. For example, before 1993, a tax differentiation was introduced in the Netherlands between cars that met the 1993 EU standards and cars that did not. Because of this differentiation, most gasoline cars were fitted with three-way catalytic converters, even in 1991 and 1992. Tax differentiation on the purchase tax for new cars has also been introduced in Sweden. More highly polluting vehicles are taxed at a higher rate than less polluting ones. This example shows how tax differentiation can be combined with emission standards: the standards set the maximum emission levels, the differentiation encourages people to buy cars that have lower emissions than the current maximum level. Such tax differentiations are also useful for encouraging a faster uptake of new technology (Bailey, 1994). One problem with tax differentiation related to emission factors of new vehicles is that the emission factors of most pollutants change (increase) over time. This increase differs per vehicle type and even per vehicle. This problem can partly be solved if a standard test is developed and used to estimate emissions for a total distance of e.g. 100,000 miles (cars) or more (trucks, buses). An alternative is to base tax differentiation for new cars on their fuel efficiency. This option is currently being investigated by the Dutch government.

Variabilization - The variabilization discussion mainly focuses on car use, although the concept can also be used for other vehicle (transport) types. There are two opinions on what variabilization is: (a) variabilization is a shift from fixed costs to variable costs, leaving overall costs (more or less) constant, (b) variabilization is any change in cost structure that results in a higher percentage of variable costs (and so in a reduction of the percentage of fixed costs). So, raising variable costs (for example by raising levies on fuels) without reducing fixed costs (for example taxes on new cars or annual taxes) is variabilization according to definition (b) but not according to definition (a). In policy discussions description (a) is the most common. In this chapter both alternatives are used.

The general idea of variabilization is that higher variable costs result in a lower level of car use, while higher fuel prices, more specifically, stimulate people to buy more fuel-efficient cars and to practice a more fuel-efficient way of driving. Even if total costs remain unchanged a shift from fixed to variable costs can reduce car use because of consumers' reactions to (changes in) costs. Boose and Van Wee (1995) analyze numerous simulations with a car use and ownership model for the Netherlands. In one simulation the taxes on motor fuels are raised by approximately 0.4 ECUs and fixed taxes are abolished in 1995, implying that total costs for consumers and government revenue would remain the same if there were no consumer reactions. The effects of this intervention in 2010 are: (a) a reduction in car ownership and car use of approximately 4 percent; (b) a reduction in fuel consumption and CO_2 emissions of approximately 6 percent; (c) an increase in NO_x emissions of approximately 30 percent and a decrease of VOCs of 25 percent (NO_x and VOC emission changes are mainly due to a shift towards diesel); and (d) a loss of government revenues from levies on car use and car ownership of approximately 11 percent.

Some general results of the simulations are: (a) the specific design of variabilization (e.g. a similar increase in levies for all fuel types or a differentiation) strongly influences the results; (b) variabilization influences car use more than car ownership unless a shift toward cheaper fuels is effected (as in the example above); (c) variabilization will likely reduce fuel consumption more than car kilometrage because of a shift to more fuel-efficient cars; (d) variabilization can result in a shift between fuel types; if the percentage of diesel cars increases, NO_x emissions will increase; (e) total revenue for the government can change significantly, depending on the design of variabilization measures; (f) car ownership will grow if it becomes cheaper, more or less compensating for the reduction in the average yearly kilometrage of cars and the shift to more fuel-efficient cars; (g) the maximum effects on car use and car ownership will be reached in 5 to 10 years. After 10 years these effects will diminish. The maximum effects on fuel efficiency will be reached in the longer term (15-20 years).

It follows that a variabilization policy has to be designed very carefully. National differences in the existing tax structure and supply of fuels can greatly influence the effects of variabilization (Bleijenberg, 1989). It is not within the scope of this chapter to discuss the most favorable way of variabilization for each EU country. Secondly, it follows that if fixed costs are reduced, the negative effects of higher car ownership levels can - under certain circumstances - more or less offset the positive effects. The third and main conclusion is that - if carefully designed - variabilization can have significant effects, especially if variable costs are raised more than fixed costs are lowered (referring to our second definition).

Subsidies - One of the main advantages of subsidies is a political one. Subsidies involve a transfer of resources from the general, diffuse population of tax payers to a very visible and identifiable group of beneficiaries (e.g. public transport users). They seldom affect any particular interest group. This, to a large extent, explains why subsidies are so widely used, in the transport sector and elsewhere. In recent years, however, subsidies have received less attention, mainly because of the economic climate. The economics of subsidies for transport is relatively complicated (Button, 1993). Some objections are that granting subsidies may cause economic inefficiencies and may lead to violation of the Polluter Pays Principle (UNECE, 1994).

Recently, in some countries subsidies have been introduced or been considered in order to stimulate the removal of old cars (without catalytic converters) from the car fleet, in order to reduce exhaust emissions of NO_x, VOC, CO and CO_2 and to improve energy efficiency. A life cycle analysis is the best way to analyze the environmental impact of such subsidies. Calculations for the Netherlands show that, at least for CO_2 and energy use, scrapping old cars is ineffective. Life cycle energy use is even *increased*, because the extra energy needed for producing cars which replace the prematurely scrapped ones, is more than the reduction of energy use by driving (newer) cars. Scrapping cars without catalytic converters may reduce life cycle emissions of NO_x, VOC and CO, but appears to be less cost-effective than fitting (unregulated) three-way converters on these cars (Van Wee and Meurs, 1994). This example does not imply that subsidies are necessarily ineffective from an environmental point of view, but only that they have to be designed carefully.

If there is no extra charge on the use of relatively polluting cars, there is no price incentive for car owners to retrofit new technology. Subsidies may provide such an incentive. Subsidies can also be a part of tax differentiation, e.g. in a system in which some new car buyers have to pay while other buyers benefit, according to the emission characteristics of cars. Another example of this integration is the use of subsidies to differentiate between old types of fuels and new types that are better for the environment, since the choice of fuel type is highly dependent on fuel costs per kilometer.

Subsidies to lower tariffs for public transport are very common in western countries. Reducing car use is often mentioned as a policy reason for these subsidies. Indeed, the cross elasticity is positive: lower costs of public transport result - ceteris paribus - in less car use. But the elasticity is very low. Model simulations in the Netherlands suggest that if train fares are decreased by 1 percent, car use will decrease by 0.02 percent. Cars and public transport serve more or less different markets (Bovy *et al.*, 1990). On the other hand, price elasticity of public transport is much higher: lower tariffs result in a higher demand for public transport. Calculations for the Netherlands show that if public transport fares are lowered *on a national scale*, total energy use (and thus CO_2 emissions) for transport will increase; the energy needed for the extra demand for public transport is more than the reduction in energy for private transport (Van Wee, 1991). This conclusion is only valid for a national decrease in the tariffs. Under other circumstances the situation may be different. So, one cannot conclude that for energy reasons it is advisable to stop all subsidies for public transport. But it certainly is not true that higher subsidies generally result in less energy use and lower emissions.

In the future, subsidies could also be applied to stimulate a shift to electric vehicles, e.g. in the form of reduced taxes. Sometimes it is argued that electric vehicles only replace emissions and do not reduce them. Even in this case, however, they can be effective for improving local air quality. With regard to national emission levels, some evaluations conclude that no important reductions can be achieved by introduction of electric vehicles. In these studies, current electric vehicles (in general cars or vans) are compared with current gasoline cars. However, this conclusion does not seem realistic. Much more research has been done to improve gasoline cars than electric cars. So, if policies change in favor of electric cars, significant improvements may occur and emission reductions could be significant. To date, the energy efficiency of electric cars is already much higher than conventional cars. Power plants can be improved as well. If, for example, heat and power generation is combined, energy efficiency could roughly double. One major problem with electric cars is the low radius of action. Perhaps new battery techniques can solve this problem. If not, hybrid cars (cars with an electric and fuel energy supply) may be the solution.

Is Pricing Policy Always the Best Alternative?

A General View

Many economists believe that the most effective and efficient solution is the best. Costs of congestion, environmental damage and accidents can be seen as external costs. These have to be internalized to reach the Pareto optimum (see Chapter 1 of this volume). According to this reasoning, almost all problems can best be solved by pricing policy. The market mechanism will guarantee the best solution from an economic point of view. Although the importance of price instruments for achieving the Pareto optimum can not be underestimated, one has to consider that the usefulness of a certain policy should not only be judged on efficiency and effectiveness criteria. Other criteria, such as information need, the level of complexity of the policy, the possibilities for implementation, maintenance and enforcement, acceptability for people and firms and other political criteria, have to play a role. The most important problem is that effective and efficient instruments often turn out to be the least acceptable (Verhoef et al., 1995). In such situations politicians can choose in favor of second (or third) best solutions or try to make the first best solution more acceptable, which is not easy. Another problem is that valuation of external costs can be very complicated, especially in the case of safety and noise.

In this section efficiency, effectiveness and other criteria will be applied. If the first best solution from an effectiveness and efficiency point of view seems to pass other criteria, this solution is preferred. If the first best solution does not pass the other criteria and there is a good second best alternative, this second best alternative is to be preferred. We will also illustrate that in many cases it is relatively easy to combine instruments: they rarely conflict. Combining different measures may be preferable if there is no perfect instrument that solves the whole problem, e.g. because there are many types of vehicles and users, or because the acceptability of one instrument may increase if other measures are taken at the same time. For example, the acceptability of fiscal instruments for car use may be higher if public transport is improved. It should be borne in mind that the effectivity of combined instruments is usually smaller than the sum of effects from the separate measures. For instance, the combination of higher emission standards for vehicles and an annual environmental check will have less effect than the sum of the effects of both measures separately.

We will confine ourselves in this section to the direct effects of different instruments on the transport system and its external costs. Giving a detailed description of indirect effects on other sectors, consumer expenditures, etc., is not within the scope of this chapter, but to illustrate the relevance of such indirect effects, two examples are briefly discussed.[11] In the first example, teleworking is encouraged and people commute less. As a direct effect, energy use and emissions from traffic will be reduced, unless the reduction in the number of trips is compensated by longer home-work distances. But people working at home might use more energy than if they work at the office. If the office building has a central heating system or other people work in the same room, the reduction in energy use in the office could be marginal. Unpublished calculations of the Dutch RIVM indicate that the reduction of energy for travelling more or less equals the extra energy for heating the home. This does not mean that the government should not encourage teleworking. First of all, it might reduce congestion. Second, if teleworking is encouraged only in summer, the overall effects might very well be positive. Third, teleworking does not necessarily mean working at home; it could also mean working at central offices in residential areas. Finally, if teleworking becomes commonplace, measures that reduce energy use for heating can be taken at the office.

As a second example, imagine a government which forbids the selling of cars with an energy efficiency below certain standards. This will certainly reduce the energy use of cars. But then the money saved on car use and ownership, may be spent on other goods or services, for example, a bigger house (requiring more energy, e.g. for heating) or an extra vacation by aircraft. Extra energy used by the alternative expenditures may be more than the energy saved on car use. These examples are only given to clarify that some measures have effects on sectors other than transport, and that it is advisable to consider the overall environmental impact of such measures.

Regulation

Before discussing the most adequate policy to solve some dominant problems related to the transport sector, we will give, in this subsection, a brief outline of current practices and future prospects for regulation. Regulatory instruments are the most common alternative for pricing instruments in order to reduce external effects of traffic and transport. Here, regulating emissions, energy use, vehicle speed, driver's licences, parking and land use are reviewed. Also tradeable permits are mentioned as a regulatory instrument, although this may also be seen as a pricing instrument.

Regulation of *emissions of pollutants* from vehicles is common in most western countries. Until now most attention has been paid to regulation of emissions of new cars. The main advantage is that this regulation can be very effective (as the 1993 emission standards for new cars have proven) and relatively simple to implement if techniques to meet the standard are generally available to vehicle manufacturers. The main problem is the opposition of authorities in countries where vehicle manufacturers cannot easily meet the proposed standards. Another problem is that - according to present EU policy - making distinctions between different regions (countries) in Europe is not possible, though the environmental situations differ significantly.

Experience in the USA shows that regulation of *energy use* by cars can be very effective (the CAFE standards - see below). Regulation of energy use can be a good alternative to pricing policy if taxes have to be drastically increased to have the same effect, and strong opposition can be expected.

Speed limits can affect safety, emissions, energy use and congestion. Immers *et al.* (1994) conclude that a reduction in the maximum speed limit from 120 km/h to 100 km/h reduces energy use by approximately 19 percent per motorway kilometer. Ferguson (1994) concludes that if current speed limits were enforced, 3.1 percent of UK car emissions could be saved. If a maximum limit of 50 mph were imposed seven percent of car CO_2 emissions could be saved. In most countries safety is the main or only argument for speed regulation. The ultimate way of regulating speed is the use of *speed limiters* or, better, regulation of the maximum speed at which vehicles are physically able to drive. If emotional aspects of such regulation did not exist, the solution would be very simple: what's wrong with regulation that makes it impossible to sell cars that can drive faster than the maximum speed on motorways? This regulation is good for the environment, for energy use and safety, because it reduces both speed and speed differences. The present generation of speed limiters only regulates maximum speeds. Better results can be expected if vehicle speed can be externally limited. The main problem is acceptability: many drivers argue that speed limiters on vehicles is like castrating the driver ......

Changing the *minimum age for driver's licenses* can be very relevant for safety, energy use and emissions. Younger people have significantly higher accident rates, not only because they have less driving experience, but also because they drive less safely. Raising the age at which people are allowed to drive cars will increase safety. In addition, it will reduce car use, car ownership and related external costs. Reduced car use and ownership first of all occurs because fewer people are allowed to drive cars. An additional effect may occur because people will have more experience with alternatives which may have an effect at an older age. Nowadays many people start their working and residential careers after they get their driver's license and buy their first car. If car ownership is delayed to an age at which they start living on their own and working, the working and residential locations might be less 'car dependent', which could have an effect on mobility even after they are allowed to get their license. The main problem is the acceptability of raising the minimum age for a driver's license: people consider the present situation as a vested right. If politicians raise the age, safety has to be the main argument. If the age is raised for environmental reasons, this can have negative long-term effects in that people may come to consider the environment as their 'enemy'.

The concept of *tradeable permits* has its origin in the pollution rights concept. Under a tradeable permit system these rights are allocated by the authorities, but, once obtained, can be traded (Button 1993). The concept encourages the development and implementation of cleaner technologies. In a perfect market the ratio between economic benefits and environmental impact is optimized within the maximum emission levels the authorities have set. The UNECE (1994) concludes that emission trading for mobile source emissions seems very complex. In Chapter 7 of this book a specific proposal is worked out for a tradeable permit system to regulate the external effects of transito traffic through the Alpine region.

Parking regulations can be used to influence land use, to reduce car use, emissions and noise, and to reduce congestion. They have been popular in western countries for many years, especially in city centres and towns. The main reason for their introduction is that they can easily be implemented (as far as they are not on private territory) and effectively solve the problems of parking and car use in central areas. Economists have argued that this kind of regulation is - from an economic point of view - less attractive than pricing parking places (Verhoef *et al*, 1995b), but the acceptability of very high fees may be a major problem.

Land use regulations can influence car use, the use of public and other modes of transportation, and so external costs of transport. It is not within the scope of this chapter to describe the complicated interactions between land use and transport. Land use regulations are also used to reduce noise nuisance. In many countries these days, it is forbidden to build close to motorways, other main roads, rail lines and airports, unless noise screens or other barriers are built between the infrastructure and buildings. Building of new infrastructures close to residential areas is also forbidden by land use regulations.

Which Policy for Which Problem?

Congestion on motorways, energy use and CO_2 emissions, emission of other pollutants, noise nuisance, and liveability of city centers can be considered as the five main external effects, caused by traffic and transport. Building on the analysis in this chapter, we will outline a synthetic view of the best (combinations of) instruments to tackle these problems.

Congestion - In general the best way to reduce congestion is road pricing, because of its potential to differentiate between locations and times. If a complex system with different prices for different periods of the day (according to congestion levels) is used, the users need to know the prices in advance (before making decisions on whether or not and how to go, and on route and time of day). The most important problem is acceptability. Second best solutions can be higher fuel prices, traffic management and parking policy at the destination (see Verhoef *et al.*, 1995a and 1995b, for the welfare economic characteristics of second best alternatives to road pricing). Finally, building new infrastructures can reduce congestion. McLaren and Higman (1992) conclude that to relieve congestion simple capacity increases are of least value, while demand management techniques are most likely to be useful. This leads to the conclusion that building new infrastructures will often be the third best solution.

Energy use and CO_2 emissions - These can be pushed back by reducing energy use and CO_2 emissions per kilometer and by reducing the amount of kilometers. Theoretically, higher levies on fuels is the first best solution for reducing CO_2 emissions. If all external costs are internalized by raising fuel taxes, standards for fuel efficiency are not needed and not effective. But acceptability of much higher levies on fuels is limited. Therefore standards for energy use or CO_2 emission standards for manufacturers can, as a second best solution, fill the gap between what is needed to internalize external costs and what can be achieved by higher levies on fuels.

Higher fuel prices directly reduce both car kilometrage (see above) and the emissions per kilometer. Reduced emissions per kilometer are the result of several processes. First, people buy more fuel-efficient cars if fuel prices are higher. Additionally, people may adopt a more fuel-efficient way of driving (lower speed, acceleration and deceleration behavior). Finally, car manufacturers may develop and offer cars with a better fuel efficiency.

In the USA, the CAFE standards have proven to be successful, although they have been criticized. The standards require the domestically produced and imported vehicles sold by each manufacturer to achieve a specified average fuel efficiency rating. In the EU the situation is slightly different, because of the much higher taxes on fuels and hence fuel prices, and the better fuel efficiency of cars. A higher tax on fossil fuels will therefore have more impact on energy use in the USA than in Europe. Speed limiters can be very effective for reducing energy use and CO_2 emissions as well. Nevertheless, improved safety seems to be a better political argument in favor of speed limiters.

Emissions of pollutants - Again, these can be decreased by reducing the emissions per kilometer and by reducing the number of kilometers. Present and expected future technical options for reducing emissions per kilometer promise further drastic reductions of emissions of cars and vans, and (to a lesser extent) of trucks. In recent years attention for other vehicle types has grown (e.g. ships, aircraft, motorcycles). Although only a few useful references were found, it seems that reducing emissions of other vehicles can be even more cost effective than reducing emissions of cars and trucks.

Theoretically, pricing emissions may be the best solution for achieving optimal emissions per kilometer. The price can be set on new cars and/or on car use. In the first case taxes on new cars are lower if cars are cleaner. Yet, pricing does not seem to be an acceptable solution for various reasons. First, there must be agreement on how to valuate different pollutants. This is difficult. Second, emissions depend on (among other things) the percentage of cold starts and driving speed. Therefore a standard test cycle has to be defined, but even then the problem is that individual driving conditions can vary considerably (e.g. depending on the percentage of cold engine kilometers). Third, emission factors of cars can change during the lifetime of vehicles. Fourth, in the case of pricing instruments related to car use, corruption can be expected and a very complex and expensive system of implementation and enforcement is needed. Therefore regulation (emissions standards) as a second best solution is to be preferred. Regulation is relatively simple and effective if the standards are high enough. Nevertheless, pricing can play an additional role. If technical possibilities are better than current standards, tax differentiation can encourage the supply and demand for new cars that perform better than the current standards. As stated before, this kind of differentiation has proven successful (e.g. in the Netherlands more than 80 percent of all new gasoline cars sold in 1992 met the EU standards, to be introduced in 1993).

Another possibility for reducing emissions per kilometer can be the use of alternative fuels, especially for (heavy) diesel vehicles. Until now the discussion has focused mainly on buses, because reducing their emissions can improve local air quality significantly. If changes in fuels for buses is a local policy, using alternative fuels will be confined to special areas (urban areas). Then, regulation could be the solution. However, given that in most European cities and towns only one or two companies are involved, voluntary agreements with these companies can also have advantages. Sharing the extra costs can be a solution to make these instruments more acceptable. National authorities can use tax differentiations to encourage companies to buy 'clean' buses, e.g. differences in the purchase tax on new buses, the yearly taxes, or the taxes on fuels. This tax differentiation instrument can also be used to change the choice of fuels for other vehicles or to promote electric vehicles.

Noise nuisance - This external effect can best be reduced by regulation. The first best solution is a combination of several measures, of which reducing noise emissions from vehicles is probably the most important. Other possibilities are building screens or other barriers between infrastructures and houses, urban traffic circulation measures, building low traffic areas or traffic-free cities and towns, using special 'open structure' asphalt and reducing traffic speed. Land use planning, both for infrastructures and dwelling locations, can help to prevent future problems. Of course, an overall reduction in the use of vehicles reduces traffic noise. Pricing instruments can be used to generate this overall reduction. But strong reductions are needed to have a significant effect on noise nuisance. In general, noise isolation of existing houses is a second best solution, because if windows are open or people sit in their gardens, noise is still a problem. Architects can also reduce noise nuisance (e.g. it is better to have the bathroom or the kitchen on the street side instead of a bedroom or living room).

Liveability - In areas like city centres liveability is sometimes reduced because of road traffic. The term liveability is used because the quality of staying in these areas is affected by numerous problems related to traffic, like noise, air quality, odor, physical nuisance of driving cars, parking and less attractive scenery. Liveability can be improved by implementing policies for these different problems, but also by a strategy to reduce overall road traffic in these areas. Then, the best solution may be a combination of access charges, reducing parking places, reducing the share of on-street parking, higher parking tariffs, and a redesign of streets and squares.

For a specific policy on problems related to *parking*, pricing instruments are the first best solution, mainly because they differentiate between locations and times of day. For an adequate parking policy the right balance between the number of parking places and tariffs has to be found. An alternative would be traffic management.

Conclusions

- Traffic significantly contributes to many environmental problems. The most important problems may be climate change, acidification, noise nuisance and local air pollution.
- To reduce the environmental impact of transport there are several instruments available. Fiscal instruments are only one of many kinds of instruments, but can play an important role in approximating the Pareto optimum.
- Many goals are relevant for transport policy. The main goal is (a) to allow people to optimize the fulfillment of their location-related needs by providing options for mobility and (b) to optimize the fulfillment of the needs for transporting goods. From an economic point of view, the transport system should be improved (changed) until marginal social benefits equal marginal social costs.
- Until recently, the main reason for pricing policy was to generate funds. Nowadays internalizing external costs of congestion and environmental damage play an additional role.
- Transport generates significant external costs. Costs of accidents, congestion and environmental damage are the most important.
- The EU transport policy on emission standards of cars and other road vehicles, and on fuels has been effective in the past and will likely be effective in future. Progress on other policies has been limited. The present policy suggests that other policies, such as a reduction of external costs by pricing instruments, will become more important in future.
- The most promising pricing instruments are higher levies on fuels, road pricing, variabilization, tax differentiation and parking fees. At least theoretically, pricing instruments are the first best solution for most problems if the Pareto optimum is to be reached. Unfortunately, acceptability of pricing instruments is very often limited. In general, the most effective and efficient instruments are the least acceptable. To have a significant effect, high taxes are needed, because elasticities are relatively low. This may be a major problem for acceptability.
- The EU can use pricing instruments to reduce external costs of transport more than it has so far. An active role can be played with regard to fuel taxes (setting the minimum levels including at least a 'consensus' level of external costs and maybe also maximum levels),

taxes on new cars and annual recurrent taxes. The EU can also provide information for individual members to help them set the desired level of fuel taxes and taxes on fixed costs, mainly based on external costs and how to price them. The EU can stimulate individual members to introduce road pricing. The EU can also provide information or guidelines on where to introduce road pricing, and what prices to set at what time of the day for different vehicle types. If new roads are (partly) financed by the EU, the Union can play an active role in introducing road pricing. The EU can increase the possibilities for individual Member States to use tax differentiation as an instrument to stimulate technologies or fuels that perform better than the standards.

- Pricing instruments (alone or combined with other instruments) are the first best solution for congestion (first best: electronic road pricing), for energy use and CO_2 emissions (fuel prices), fuel and vehicle related emissions (tax differentiation), and parking problems (fees).
- For some problems, such as emissions of pollutants per vehicle kilometer and noise nuisance, pricing instruments might theoretically be the first best option, but implementing them is so complicated that regulation is the preferred solution.
- In general, specific pricing instruments can mostly be used in combination with other pricing instruments and regulation; they seldom exclude each other.
- Indirect effects of transport policy on other sectors and on consumer expenditures should be taken into account, in order to evaluate overall effects.

Notes

[1] The author thanks E. Verhoef (Free University, Amsterdam) for his comments on the draft version of this chapter and H. Geerlings (Erasmus University, Rotterdam) for providing some literature on EU policy.

[2] Transport refers to the transport of people and goods, traffic refers to vehicle use on roads, rail and water and in the air. In general transport volumes are expressed in passenger or ton kilometers, traffic volumes in vehicle kilometers.

[3] The application of pricing instruments can be judged on efficiency, effectiveness and equity criteria. For a discussion on this subject related to transport, see Verhoef (1993) and Verhoef, Nijkamp and Rietveld (1995).

[4] Based on expected EU policy; calculations made for the Netherlands.

[5] The target is a reduction of 10% between 1986 and 2010. Between 1986 and 1990, however, the CO_2 emissions increased.

[6] The most common methods are: a) direct monetary costs, b) indirect monetary costs, c) avoidance costs, d) willingness to pay, e) willingness to accept. It is beyond the scope of this chapter to describe these methods.

[7] Legislation dates from before 1993, therefore this subject is discussed in this section.

[8] For information about transport policy and the environment in different countries, see Barde and Button (1990).

[9] See Chapter 3 of this volume on the recent EU proposal to introduce a carbon/energy tax. This additional tax on fuels would have a hybrid tax base, i.e. a combination of an excise duty on the carbon content of fossil fuels and an excise duty on all non-renewable forms of energy.

[10] This section is based on Verhoef *et al.*, 1995a.

[11] For more information on this subject see Verhoef and Van den Bergh, 1994.

References

Bailey, P. (1994), *Economic Instruments to Control Transboundary Air Pollutants from Road Transport in Europe: A Discussion,* paper presented to the Task Force of Economic Aspects of Abatement Strategies, UN/ECE, Geneva.

Barde, J.P. and K. Button (Eds.) (1990), *Transport Policy and the Environment. Six case studies,* Earthscan Publications Ltd., London.

Bleijenberg, A.N. (1989), *European Variabilization of Car Costs,* Center for Energy Conservation and Clean Technology, Delft (in Dutch).

Bleijenberg, A.N., J.W. Velthuijsen, T. Oegema (1990), *Economic Consequences of (Auto)Mobility Control,* CE/SEO, Delft/Amsterdam.

Blok, P.M. and G.P. van Wee (1994), 'The Traffic Question'. In: F. Dietz, W. Hafkamp and J. van der Straaten, *Handbook Environmental Economics,* Boom, Amsterdam/Meppel, pp. 216-234 (in Dutch).

Boneschansker, E., M.G. Lijesen, H. de Groot (1994), *The Price of Mobility in 1990. Part 1: Summary,* Government Expenditure Research Institute (IOO), The Hague.

Boose, J.J.E.C., G.P. van Wee (1995), *Influence of Changes in Incomes, Costs and Speeds on Car Ownership, Car Use, Energy Use and Emissions. Results of 151 Simulations with FACTS 2.0,* National Institute of Public Health and Environmental Protection (RIVM), Bilthoven (in Dutch).

Bovy, P.H.L., J. van der Waard, A. Baanders (1991), *Substitution of Travel Demand between Car and Public Transport: a Challenge to Policy Makers,* PTRC Summer Annual Meeting, Brighton.

Button, K.J. (1992), *Alternatives to Road Pricing,* paper presented to the OECD/ECMT/NFP/GVF Conference on 'The Use of Economic Instruments in Urban Travel Management', Basel.

Button, K. (1993), *Transport, the Environment and Economic Policy,* Edward Elgar, Aldershot.

Commission of the European Communities (CEC) (1990), *Energy in Europe. Energy for a New Century: The European Perspective,* Office for Official Publications of the European Communities, Luxembourg.

CEC (1992a), *Towards Sustainability. A European Community Programme of Policy and Action in Relation to the Environment and Sustainable Development,* Brussels.

CEC (1992b), Green Paper on *The Impact of Transport on the Environment. A Community Strategy for Sustainable Mobility.* + White Paper on *The Future Development of the Common Transport Policy. A Global Approach to the Construction of a Community Framework for Sustainable Mobility,* Brussels.

CEC (1993), *Growth, Competitiveness, Employment. The Challenges and Ways Forward into the 21st Century,* White Paper, Brussels.

Evans, A.W. (1992), 'Road Congestion Pricing: When Is It a Good Policy?', *Journal of Political Economy,* Vol. 100, Nr. 1, pp. 211-217.

Ferguson, M. (1994), 'The Effect of Vehicle Speeds on Emissions', *Energy Policy,* February.

Goodwin, P.B (1992), 'A Review of Demand Elasticities with Special Reference to Short and Long Run Effects of Price Changes', *Journal of Transport Economics and Policy,* pp. 155-169.

Group Transport 2000 Plus (1990), *Transport in a Fast Changing Europe,* Brussels.

Immers, L.H., G.R.M. Jansen, R.C. Rijkeboer (1994), *Reducing Fuel Consumption of Cars: A Research Agenda,* TNO, Delft (in Dutch).

Kageson, P. (1993), *Getting the Prices Right. A European Scheme for Making Transport Pay its True Costs,* Katarinatryck AB, Stockholm.

Knight, F.H. (1924), 'Some Fallacies in the Interpretation of Social Cost', *Quarterly Journal of Economics,* Vol. 38, pp. 582-606.

Lovins, A.B., J.W. Barnett, L. Hunter Lovins (1993), *Supercars. The Coming Light-Vehicle Revolution,* Paper presented at the Summer School of the European Council for an Energy-Efficient Economy, Rungestedgard, Denmark.

McLaren, D. and R. Higman (1992), *The Environmental Implications of Congestion of the Interurban Network in the UK,* Paper presented at the PTRC summer annual meeting.

MacKenzie, J.J., R. C. Dower, D.D.T. Chef (1992), *The Going Rate: What it Really Costs to Drive,* World Resources Institute, Washington.

MacKenzie, J.J. and M.P. Walsh (1990), *Driving Force: Motor Vehicle Trends and their Implications for Global Warming, Energy Strategies, and Transportation Planning,* World Resources Institute, Washington.

NEA (1991), *Congestion Costs on the Dutch Motorways in 1990,* NEA, Rijswijk.

NEI (1991), *Price Elasticity of Energy Use in Road Traffic,* Netherlands Economics Institute, Rotterdam (in Dutch).

Organization for Economic Cooperation and Development (OECD) (1988), *Transport and the Environment,* OECD, Paris.

OECD (1993a), *Cars and Climate Change,* OECD/IEA, Paris.

OECD/IEA (1993), *Energy Prices and Taxes,* Nr. 1.

Pigou, A.C. (1920), *Wealth and Welfare,* Macmillan, London.

Quinet, E. (1990), *The Social Costs of Land Transport,* OECD, Paris.

RIVM (National Institute of Public Health and Environmental Protection) (1992), *The Environment in Europe: a Global Perspective,* RIVM, Bilthoven.

RIVM (1994), *National Environmental Outlook 3, 1993-2015,* Samson H.D. Tjeenk Willink, Alphen aan den Rijn (Dutch version: 1993).

Rothengatter, W. (1993), 'Externalities of Transport'. In: J. Polak and A. Heertje (Eds.), *European Transport Economics,* Blackwell, Oxford, pp. 81-121.

Rotterdam Transport Centre and Erasmus Centre for Environmental Studies (1991), *Transport and the Environment. Interim Report SAST Project-3,* RTC/ESM, Rotterdam.

Smith, R.S. (1994), *Taxes on Motor Vehicles and Their Use in the European Community -*

Environmentally Sensitive Trends?, paper presented at the 50th Congress of the International Institute of Public Finance, 'Public Finance, Environment and Natural Resources', August 1994.

UN, Economic and Social Council (1990), *The Effectiveness of Economic and Other Instruments to Reduce Air Pollution Emissions from the Transportation Sector,* Report by the Secretariat.

UNECE (1994), *Economic Instruments to Reduce Mobile Source Emissions. Potentials and Experiences,* Secretariat of the United Nations Economic Commission for Europe, with the assistance of J. Baijens (preliminary draft).

Van Gent, H.A. and P. Rietveld (1990), 'Road Transport and the Environment in Europe', *Milieu,* Vol. 5, nr. 6.

Van Wee, G.P. (1991), 'Better Public Transport, Better Environment?', *Verkeerskunde,* Nr. 7/8, pp. 14-15 (in Dutch).

Van Wee, B. and H. Meurs (1994), 'Shortening the Age of Cars: Good or Bad for the Environment?' In: J.M. Jager (ed.), Colloquium Vervoersplanologisch Speurwerk 1994, C.V.S., Delft (in Dutch).

Van Wee, G.P., J. van der Waard, M.J. van Doesburg, H.C. Eerens, H. Flikkema, A.L. 't Hoen, E. Rab, R. Thomas (1993), *Traffic and Transport in the Dutch National Environmental Outlooks and the TSTP-outlook 1993,* RIVM/NEA, Bilthoven/Rotterdam (in Dutch).

Verhoef, E.T. (1993), 'External Effects of Road Transport: Some Theory and a Survey of Empirical Results', TRACE discussion paper TI 93-35, Tinbergen Institute, Amsterdam-Rotterdam (forthcoming in *Transportation Research* 28A:3).

Verhoef, E.T. (1994), 'Efficiency and Equity in Externalities: A Partial Equilibrium Analysis', *Environment and Planning A,* Vol. 26, pp. 361-382.

Verhoef, E. and J. van den Bergh (1994), *Transport and Environmental Sustainability. An Adapted SPE Approach for Modelling Interactions between Transport, Infrastructure, Economy and Environment,* TRACE discussion paper TI 94-63, Tinbergen Institute, Amsterdam.

Verhoef, E., P. Nijkamp, P. Rietveld (1995), *The Trade-off between Efficiency, Effectiveness, and Social Feasibility of Regulating Road Transport Externalities,* TRACE discussion paper TI 95-11, Tinbergen Institute, Amsterdam.

Verhoef, E., P. Nijkamp, P. Rietveld (1995a), 'Second-best Regulation of Road Transport Externalities: the Case of Regulatory Parking Policies', *Journal of Transport Economics and Policy,* forthcoming.

Verhoef, E., P. Nijkamp, P. Rietveld (1995b), 'The Economics of Regulatory Parking Policies', *Transportation Research,* forthcoming.

Vollebergh, H. (1994), 'Transaction Costs and European Carbon Tax Design'. In: M. Faure, J. Vervaele and A. Weale, *Environmental Standards in the European Union in an Interdisciplinary Framework,* MAKLU, Antwerpen, pp. 135-154.

Vougias, S., (1992), 'Transport and Environmental Policies in the EC', *Transport Reviews,* Vol 12, Nr. 3.

Walsh, M.P. (1990), *The Impact of Transport on Health and the Environment,* paper prepared for the World Health Organization, Commission on Health and the Environment.

Whitelegg, J., A. Gatrell, P. Naumann (1993), *Traffic and Health,* Environmental Epidemiology Research Unit, University of Lancaster, Lancaster.

Zahavi, Y. (1979), *The UMOT Project,* US Department of Transportation, Report DOT-RSPA-DPB-2-79-3, Washington.

7 Tradeable Permits for Transito Traffic through Austria

Paul R. Koutstaal

Introduction

Since the establishment of the European Communities - now the European Union (EU) -, the economies of the Member States have grown considerably. This growth has been accompanied (and to a certain extent caused) by a much higher growth in the internal trade of the community. Figure 7.1 shows that internal trade of the (then) 12 states of the European Union increased tenfold between 1970 and 1992, while real GDP barely doubled. More trade means more transport and traffic, and although trade in itself is attractive from an economic point of view, the increase in traffic can be a burden for the environment. This has become particularly clear in two European countries, Austria and Switzerland, who bear the brunt of the transito traffic between the northern and the southern part of the European Union. Transit traffic through Austria has risen considerably in the period 1970-1989, as can be seen in Figure 7.1 which shows the total weight of goods transported through Austria by means of trucks. Given an average payload per truck, this graph is representative of the growth in the number of movements. The dip in 1990 was caused by the ban on night driving of trucks introduced in Austria at the end of 1989, the closure of the Kufstein motor bridge and long lasting-strikes on the Italian border which obstructed freight traffic.

Why is this traffic an environmental burden? One of the reasons is that the transito traffic is concentrated on a few major routes, like the Brenner route through Tirol in Austria. Consequently, the effects of this traffic are felt heavily in relatively small areas which, moreover, are environmentally vulnerable mountain areas. The environmental damage done by traffic in these areas is diverse. The major problems are caused by the emissions of carbon monoxide (CO), hydrocarbons (HC), particles and nitrogen oxides (NO_x), which damage both health and the alpine environment. Apart from its effect on local health, NO_x emissions contribute to acid rain.[1] The pine forests of the Alps have suffered heavily from acid rain. Other aspects of the environmental burden of traffic are noise and vibration, the use of land for road infrastructures (which is especially a problem in the mountain valleys where usable land surface is scarce) and the destruction of wildlife habitats.

In order to reduce the burden of this transito traffic, from which they mainly get the costs and not the benefits (because the traffic is transit traffic, there is hardly any offspin to the Austrian and Swiss national economies), Austria and Switzerland have negotiated transit agreements with the European Union (CEC, 1992; 1993).

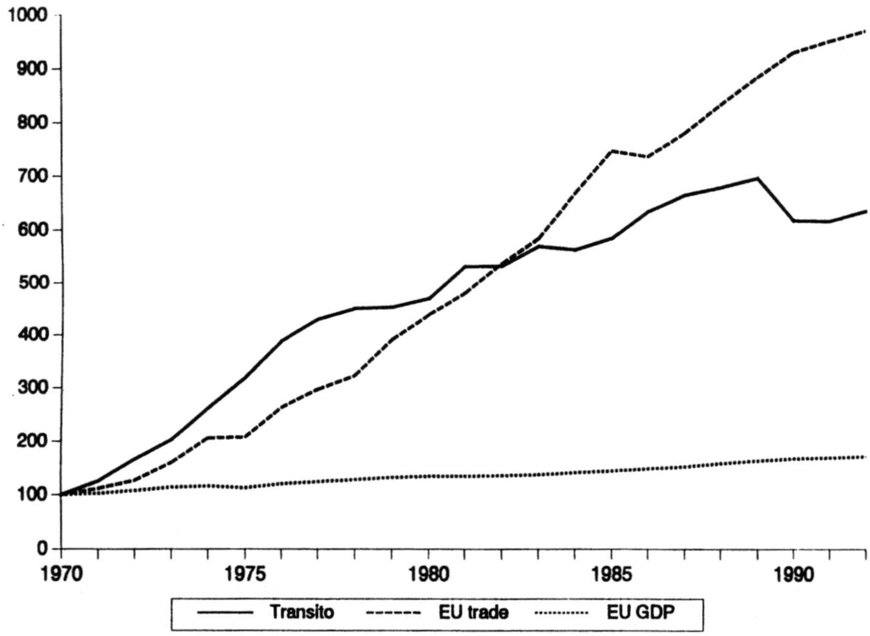

Figure 7.1: GDP and internal trade of the European Union, and transito traffic through Austria (1970 = 100).

In these transit agreements, both the economic considerations of the Union and the environmental concerns of the Alpine countries are taken into account. Apart from this similarity, the agreements with Austria and Switzerland differ considerably in the approach used to reduce the burden of transit traffic. Transito traffic through Switzerland was already severely limited because no trucks over 7.5 tons are allowed through. For this reason, the agreement between Switzerland and the EU focuses on increasing the infrastructure for combined traffic. The agreement with Austria however, focuses on the reduction of NO_x emissions (which, according to the agreement, will be technically the most difficult task as compared with reducing the other harmful emissions) and if necessary, limitation of the number of transit movements. In order to reach this goal, the agreement with Austria introduces a system of 'ecopoints', which can be seen as a permit system for transito traffic through Austria.

Both agreements came into force on the first of January, 1993. At that time neither country was a member of the European Union. Since then Switzerland has decided to stay outside the Union, Austria however entered the EU in January 1995, as did Finland and Sweden. For this reason the transit agreement with Austria has been renegotiated and reaffirmed as an internal EU regulation for the now 15 Member States of the Union (CEC, 1994).

In this chapter we analyze the effectiveness and efficiency of the ecopoint system, introduced in the agreement between Austria and the EU and nowadays in force as an EU regulation, and we evaluate the possibility of improving efficiency by making the ecopoints tradeable, thus enlarging the flexibility of the system. In the next section, the transit agreement between the EU and Austria and its recent changes will be described. Then, the regulatory instruments used to reduce pollution and the environmental burden of the transit traffic are evaluated. A major shortcoming of the agreement and the succeeding regulation is the lack of tradeability of the ecopoints. As a result, EU-wide abatement costs will be higher than is necessary. The fourth section will discuss how the agreement could be improved in order to make it work more efficiently by making use of a more market-oriented approach. Next, a rough estimate is given of the potential for trade between different countries and within countries. Finally, it will be argued that the proposed approach is not only efficient but also politically attractive and in agreement with the spirit of the Common Market.

The Austria-EU Transit Agreement: A Command-and-Control Approach

Environmental Objectives

The purpose of the transit agreement between Austria and the EU is to limit the environmental burden of transit traffic originating from Member States of the European Union. As has been described above, this traffic causes several environmental problems.

The reduction of NO_x emissions is to be achieved by a system of so-called 'ecopoints'. Each truck of over 7.5 tons which passes through Austria must deliver a certain number of ecopoints. The number of points depends on the average NO_x emissions of the truck. The more NO_x it emits, the more points it needs. Each truck must have a document which contains its 'Conformity of Production' (COP) emission level. In this document the emission level of the truck is determined in grams NO_x per kWh of the truck (a kWh is a measure of energy; the characteristic emission level could also have been expressed in grams NO_x per liter of fuel used). For each gram NO_x per kWh the truck emits, one ecopoint must be 'paid', each time the truck passes through Austria. For trucks which do not have a COP certificate (for example trucks registered before 1 October 1990) the COP level is set at 15.8 g/kWh. The ecopoints are valid only for one year, so it is not possible to accumulate them for use in another year.

Basically, the ecopoint system is a command-and-control system, where each polluter needs a permit in order to be allowed to emit. An important difference with the usual approach however is that in the ecopoint system the total amount of emissions is limited by the number of points which is made available. In the standard regulatory approach, a firm acquires a permit if it abides by the rules set by the authorities for its type of pollution and activity.[2] Therefore, the total amount of pollution is not restricted: when new polluters enter, total emissions will rise.

In the original agreement between the EU and Austria, the total number of ecopoints made available for trucks from the European Union was based on the number of runs by European trucks through Austria in 1991. This number, 1,264,000 runs, multiplied by 15.8 (the COP level for older trucks) gives the total amount of ecopoints in the imaginary base year, 1991. Therefore, in 1991 about 20 million ecopoints would have been available for trucks from the EU. In subsequent years, the number of ecopoints would be reduced, as can be seen in Table 7.1. For example in 1993, the first year that the treaty was in force, the total amount of ecopoints was already 12.1% less.

Table 7.1: Amount of ecopoints relative to base year (in %).

1991	*100.0*	1996	*65.0*	2000	*49.8*
1992	*96.1*	1997	*59.1*	2001	*48.5*
1993	*87.9*	1998	*54.8*	2002	*44.8*
1994	*79.5*	1999	*51.9*	2003	*40.0*
1995	*71.7*				

In 2003, the number of ecopoints and therefore total NO_x emissions will be reduced by 60 percent. There is one condition in the agreement: in setting the reduction percentages for the last two years, it is assumed that in these years trucks are available with a COP level of 5.0 g NO_x/kWh (current emission standards are 8.0 g NO_x/kWh). The treaty does not mention what will happen if this is not the case.

Under the ecopoint system, NO_x emissions are reduced considerably. However, this does not guarantee that the number of transit movements is reduced as well. Consequently the environmental problems which are caused especially by the number of runs, like noise and vibration and the use of land, will not necessarily be reduced. If the average COP level is lowered quickly by the transport firms, the situation might occur that more runs are made without exceeding the intended level of NO_x emissions, because fewer points are needed per run (the truck is cleaner). Therefore there is a special clause in the agreement, which is intended to slow down the number of runs. If in one year the number of runs exceeds the 1991 level by 8 percent, the amount of points for the next year is determined by a special rule instead of by the reduction path in Table 7.1. The NO_x level of the current year is extrapolated to determine the average level for next year. This is multiplied by 1.04 and by the number of runs in the base year, 1991. Compared with the year before in which total runs exceeded the 1991 level by 8 percent or more, the number of ecopoints is reduced at a quicker rate than is indicated in Table 7.1. This leaves the transport firms with two options. They may further reduce the COP level of their trucks, or they have to limit the number of runs. In the latter case the ecopoint system not only reduces the NO_x emissions, but to a certain extent also limits other environmental problems which are connected with the number of runs. It is obvious however, that the system does not impose an absolute restriction on the volume of transito traffic.

Three final remarks should be made. First, several kinds of transit traffic are exempted from the ecopoint system, e.g. postal traffic, traffic for cultural activities like stage plays and sports activities, and transport related to calamities. Second, the agreement also contains several articles on investments in infrastructure for combined transport of goods through Austria (transport of trucks and goods by train). Various projects like freight handling facilities, tunnels and railways which both parties to the agreement should undertake are specified. Third, apart from the ecopoint system, both parties in the agreement have pledged to reduce the emission standards trucks have to comply with. The emission standards for CO, HC, NO_x and particles are to be decreased by October 1, 1996. As is stated in the agreement, these standards will be based on 'the most advanced and economically attractive technology'.

The Entry of Austria into the EU[3]

The agreement between Austria and the EU had to be renegotiated after Austria applied for membership in the European Union. As could be expected, the future of the transit agreement has been a difficult point in the negotiations. The opening bid of the EU was to maintain the agreement for only three years after the entry of Austria in the Union. This was unacceptable for Austria because it would leave transit traffic unhampered after this transitional period, with dire consequences for the environment. Apart from the environmental concerns, the Austrian negotiators had to bear in mind the public opinion in their own country. Joining the EU had to be endorsed by a referendum and strong resistance on this subject could thwart entry.

The end result of the negotiations was that the agreement will last for six years after Austria's entry into the Union (until January 2001). During this period, the effects of the agreement on the environment will be studied. If the targets are met in a 'durable way', the agreement will be abolished. In this case, the agreement can only be extended for another three years if the Council of the EU decides so with a qualified majority. If the development of the environment does not warrant the abolishment of the agreement, it will be extended for another three years (until January 2004). The Council can decide to abolish it anyhow, but then it needs unanimity. In fact, Austria has a veto in this last case. We may conclude that potentially, the agreement can run its full course if environmental targets are not met in a durable way. However, it is not quite clear from the agreement what is meant by this phrase. Presumably, there will be a lot of discussion about whether or not the agreement should be extended at the end of 2000.

On the first of January 1995 Austria, together with Finland and Sweden, became new Member States of the European Union. From that point the ecopoint system was in force as an internal regulation of the EU. Compared with the preceding years the system has undergone very little change, except that the total amount of ecopoints in the base year has been recalculated to allow for the participation of the new Member States. So, the number of transito runs through Austria in 1991, originating from the 15 Member States (including Austria itself), is estimated at nearly 1.5 million, implying that in this imaginary base year a total amount of 23.7 million ecopoints would have been available for the 15 countries.

How the System Works

The ecopoint system, as it now functions within the EU, is administrated in a relatively simple way. A truck passing through Austria must have a form (the 'ecoform') on which the ecopoints it needs are recorded. This form is handed in when entering Austria. The truckdriver receives a form proving that he has handed in his points. This form is given back when he leaves Austria and enters Germany or Italy, together with a copy of the COP document stating his standard NO_x emissions per unit of energy.

Each transport firm receives ecopoints from its own national authority, according to some allocation scheme chosen by this authority. The method of allotment is not laid down by the EU regulation. The practice in the Netherlands and other countries is that firms who need ecopoints receive them after applying for them. The allocation within the Member States is an administrative procedure and firms do not have to pay for the ecopoints.

The rules of distribution of the ecopoints between the Member States, however, are laid down in the regulation. According to these rules 96.66 percent of the total available number of ecopoints are distributed over the Member States in conformity with each country's share in the transito traffic through Austria in the base year, 1991 (shown in Table 7.2, second column). The remaining 3.34% are kept in reserve by the European Commission. So, the number of points which a country receives is equal to its share in the base year, times the available number of points in a given year minus the reserve. Expressed as a formula:

$$P^i_t = s^i \, r_t \, T_b \, 0.9666$$

in which:

P^i_t is the number of points for country i in year t
s^i is the share of country i (shown in Table 7.2)
r_t is the reduction factor in year t (shown in Table 7.1)
T_b is the total number of points available in the base year

Two countries have been accorded a special position, Italy and Greece. Of the reserve of 3.34 percent, 0.41 percent is reserved by priority for these two countries. Apart from this, the reserve is intended to alleviate Member States who have problems like a relatively disadvantageous starting point, problems with cleaning up their truck fleet and unforeseen circumstances. A Member State which does not use all its ecopoints should return its points to the European Commission (no later than the 15th of October of the year concerned), who will add them to the reserve and can subsequently distribute them to countries which have used up their permits before the end of the year.

Table 7.2: Number of runs and shares of Member States in transito traffic through Austria in 1991.

Member State	*number of runs*	*share (in percent)*
Austria	214,800	14.34
Belgium	32,500	2.17
Denmark	40,500	2.70
Germany	482,500	32.22
Greece	60,500	4.04
Spain	1,200	0.08
Finland	4,600	0.31
France	5,000	0.33
Ireland	1,000	0.07
Italy	510,000	34.06
Luxembourg	5,000	0.33
Netherlands	123,500	8.25
Portugal	400	0.03
Sweden	7,500	0.50
United Kingdom	8,500	0.57
Total	1,497,500	100.00

During the first year the transit agreement was in force, there were no problems in any Member State with this administrative allocation scheme. The ecopoints have not been scarce so far. In some countries there was even a substantial surplus. The reason for this is that the distribution rule is only based on national shares in total transito traffic and takes no account of differences in average emission levels. In the Netherlands, for example, about 50 percent of the ecopoints were not used. The Dutch trucks have a relatively low level of NO_x emissions. The amount of ecopoints needed for a trip is therefore small and the ecopoints are not yet scarce. In Italy and Germany, however, the number of points available initially was not sufficient. Both countries have received points from the reserve of the EU (Maurik, Dutch Ministry of Traffic, personal communication). Presumably, when the ecopoints are reduced as planned, scarcity will develop (see also below). It should be noted however that the expectation of some of the Member States is that the ecopoints will not become scarce, because of the use of combined transport and the availability of cleaner engines in the future.

It is not clear how the points will be distributed within countries should they become scarce. Given that the ecopoints are not tradeable and will not be sold, some administrative rule must be found to allocate the available points among the firms who need them.

Evaluation

The ecopoint is to some extent an effective instrument to reduce pollution. NO_x emissions will be reduced considerably. However, the number of runs is not necessarily limited and therefore the problems which are specifically caused by the number of runs, such as noise and vibration, will not decrease either.

Although the ecopoint system is partly effective, emissions are not necessarily abated at the lowest possible costs. EU transport companies who use the routes through Austria need to have ecopoints. The scarcity of ecopoints available in each Member State will determine to what extent they will need to reduce emissions and/or limit their transit runs. If ecopoints in a Member State are relatively scarce in comparison with other Member States, NO_x emissions from trucks from the first Member State will have to be reduced further in order to be able to continue transporting goods through Austria. Presumably, the marginal costs of reducing emissions will be higher when emissions have to be lower. Therefore, marginal costs of NO_x abatement of transit traffic will be higher in the Member State where the ecopoints are relatively scarce. Consequently, marginal abatement costs can differ between countries and the system is not necessarily efficient. To some extent, differences in scarcity and the connected loss of efficiency will be reduced by the administrative rule that countries which do not need all their points, should hand them in for redistribution. However, this will only function as long as there are some Member States which do not need all their points.

Another problem is that the system does not make clear to the firms to what extent they should reduce emissions. While firms will be able to estimate their need for ecopoints in the coming years, it will be difficult for them to estimate total demand for ecopoints in their Member State over the next few years. Therefore the problem arises for the firms that they do not know to what extent they must reduce their emissions so that enough points are available each year. If they guess incorrectly, they might end up not having enough ecopoints. Reducing emission levels in the short run will be impossible or very costly. Consequently, firms which can not get the points they need must use more expensive routes around Austria or more expensive modes of (combined) transport. This will increase their cost price and reduce their competitiveness.

Tradeability of Ecopoints

The shortcomings, mentioned in the last section, are basically the result of the fact that the ecopoint system in its present form does not provide adequate information about the relative scarcity of ecopoints. If, instead of command-and-control regulation, some form of economic instruments were to be used, a price mechanism would come into existence as a possibly efficient instrument to generate and disperse this information. The ecopoint system can be amended in a simple way to provide information on scarcity. The system described above has many characteristics of the instrument of tradeable emission permits, except for the tradeability of the ecopoints. A simple solution is to allow the transport firms to trade in the ecopoints.

When the points are made tradeable, it becomes attractive for firms with high abatement costs (e.g. firms from countries where the points are relatively scarce) to buy ecopoints from firms with low abatement costs (e.g firms in countries where the points are relatively abundant). The costs of reducing NO_x emissions are lower for the latter firms, therefore it is attractive for both parties to trade in the points. The firms which sell points can make a profit by reducing their emissions and selling the points they no longer need. The firms which buy ecopoints will also be better off, because for them it is cheaper to buy points than to reduce their emissions further or use longer and more expensive routes or modes of transport.

When a market has developed, the points will have a price. This price will reflect the scarcity of the points. When demand for points is large, relative to their supply, the price will rise. The market price provides the information on which the firms can base their decisions about whether they should reduce emissions further or use more ecopoints per transit movement. Of course, even though the price of the ecopoints provides information about their (future) scarcity, this does not mean that firms can predict future availability of the points exactly. The point is that without a price which signals the scarcity, estimating future demand is much more difficult. It is extremely difficult for the authorities to perform this function of the price mechanism, because they do not have the information about the costs of emission reduction of different firms at their disposal.

In a system of tradeable emission permits, pollution permits can be distributed in two ways: they can either be sold to polluters or they can be distributed for free, which is called grandfathering. The second method is probably the most attractive one in the ecopoint system, because it is the method currently employed for allocating the ecopoints to the Member States. Selling the points would be unattractive because this would increase the costs of the firms. The points can be grandfathered on the basis of the runs a particular firm made in 1991, or any other allotment rule. An attractive feature of tradeable permits is that the distribution method does not affect the efficiency of the instrument. Because of the tradeability of the points, firms will trade in ecopoints until marginal abatement costs are equal between the firms. As long as this is not the case, profitable deals can be made and trade will take place.

The same argument also applies at the European level. It does not matter for the efficiency of the instrument how many points are allotted to which country. As long as the points are freely tradeable across the borders of the Member States, emissions will be abated at the lowest possible costs. Therefore, the allocation of the ecopoints can be made in a politically expedient way. The allocation formula already used in the system (grandfathering to the Member States on the basis of the 1991 runs) can be continued, as this is evidently the politically acceptable solution.

An important condition for well-functioning tradeable permit systems is the development of a well-functioning market, which can provide a clear price signal to the transport firms. In the systems of tradeable permits that have been implemented in the past, the development of such markets has sometimes been lacking (Hahn and Hester, 1989). What can be expected of the market on tradeable ecopoints? It is difficult to predict exactly, however there are several factors which will influence an emerging market. One important factor is the number of potential actors on the market. If there is only a small number of trading parties, trades will be few and far between. Price forming will, in that case, be erratic, because there are not enough transactions for the development of a stable price. The number of potential traders in the ecopoint system is large, because all transport firms are possible actors on the ecopoint market.

Not only is it important that there are enough potential actors on the market, they should also have an incentive to trade. Apart from differences in abatement costs mentioned above, there are two more reasons why firms should be willing to trade. When the ecopoints are grandfathered on the basis of the 1991 runs, the development of the transit traffic will create differences between the firms. One firm might see its number of runs reduced and therefore have a surplus of ecopoints while another firm might need more ecopoints than it was allotted. The existence of surpluses and shortages is an incentive for trade. Trade will also develop because not all firms will invest in cleaner trucks at the same time. It is attractive for a firm to reduce NO_x emissions when it has to replace a truck. Otherwise, buying a new, cleaner truck before the old one is completely written off is more costly. Therefore the optimal point in time at which firms are willing to invest in cleaner trucks will differ between firms. The tradeability of the ecopoints allows firms to choose the optimal time of investment in cleaner trucks themselves because they can buy ecopoints from other firms and postpone the replacement of old trucks.

We want to point out one potential criticism of the tradeability, namely that more efficient transport firms from some Member States could buy permits and drive out less efficient firms from other Member States, presumably the less rich Member States. However, the idea of the European Union is that it creates one market in which firms of all states can compete with each other on equal terms. Using a system of non-tradeable transito rights which protects firms from competitors from other Member States, is contrary to this general aim of the EU.

Assessing the Potential for Trade in Ecopoints

Ideally, one would model the market for ecopoints, determine which trades would take place and estimate the cost advantage of making the ecopoints tradeable. However, it is difficult to try and calculate the potential cost advantages for a number of reasons. One would have to make estimates of the future transit movements through Austria for each Member State of the Union. While in general some expectations can be formed about future transit movements, more detailed estimates of per country development of freight traffic through Austria are difficult to make.

Furthermore, the type of trucks used for the transit traffic should be known. This will be determined by (a) the composition of the truck fleet of a given country and (b) the part of this fleet which is used on the trans-Austria route. Estimates of the future fleet per country can be made on the basis of the development of the truck fleet in past years (see Samaras and Zierock, 1992). The available data, however, do not allow us to determine which trucks in particular are used on the trans-Austria route.

Finally, abatement costs for reducing NO_x emissions from trucks should be known. Determining the (marginal) abatement costs of NO_x reduction poses difficulties. To a certain extent reduction in NO_x emissions tends to be a result of the general improvement in truck engines. Separating the costs of lower NO_x emissions from the overall costs of newer types of engines is therefore next to impossible (RIVM, 1993; Van Wee, RIVM, personal communication). When emissions have to be reduced below 4.0 g NO_x/kWh, additional measures are necessary, because it is not possible to reduce NO_x emissions further with process-integrated measures. Further abatement is possible by using SCR (Selective Catalytic Reduction). The additional costs of buying a truck with SCR which will emit less then 4.0 g/kWh, is estimated to be between 7,500 and 40,000 ECUs.

Given these problems, it is tricky to estimate the cost advantage of making the ecopoints tradeable. Instead, a limited approach is used. A rough estimate is given of the required average NO_x levels of the trucks of different Member States and their actual NO_x levels when permits are not tradeable. Such an estimate gives an idea of the potential for trade between firms from different countries. Subsequently, some other factors which influence the price of the ecopoints will be discussed.

A Guestimate of Future Required NO_x Levels per Country[4]

An important factor for determining required NO_x levels is the development of the number of runs through Austria, made by trucks from all countries involved. Some indications about the future development of transit movements through Austria can be gleaned by looking at the past. It can be estimated from the data on which Figure 7.1 is based, that, during the period 1978-1992, a rise in GDP of the European Union of 1 percent has led to an increase in the tons of goods transported through Austria by 1.71 percent before the ban on night driving.[5] Assuming that this relationship holds in the future, a yearly economic growth of, say, 2 percent would increase the weight of goods transported through Austria by about 3.4 percent each year. If the average weight of goods per truck remains equal (which is a plausible assumption), this means that the number of runs through Austria would also increase by 3.4 percent.

This result can be compared with the results of a study by NEA (1992), which estimates a yearly increase in international goods transport by road of 5.5 percent in the period 1989-2010 in Western Europe. Their estimate is based on an expected economic growth of 3.1 percent. Using the regression above, a yearly economic growth of 3.1 percent would mean that transit would increase by 5.3 percent annually. The increase in goods transported is not necessarily all by road; one alternative is combined transport, the transport of trucks or containers by rail. However, the role of combined transport appears to be limited (see NEA, 1992; topic 4).

Assuming in our *guestimate* a yearly increase of total EU transito traffic of 5 percent, the next step is to try and determine the relative growth per country. Samaras and Zierock (1992) have estimated the growth of the truck fleets for the period 1985-2000. Their estimates are based on assumptions about future population and GDP levels. It is assumed here that the development of the truck fleets of the Member States is representative of the development of their transito traffic through Austria. Table 7.3, column 2, shows the expected number of runs by trucks from each country in 2003. The next column gives the yearly growth for each country.

Given the number of available ecopoints in 2003 for each country, the required average emission level (COP level) of trucks can be calculated. The number of points each country will receive, is given by the distribution rule and by the total amount of permits available in a given year (see the second section). Column 4 in Table 7.3 shows the required average COP levels. In this calculation, no account has been taken of the special rule which applies if the total volume of transito traffic grows by 8 percent or more, relative to 1991. As has been described above, in this situation the available amount of ecopoints is reduced faster. With a growth in the number of runs of 5 percent each year, the total number of runs will soon exceed the 1991 level by more than 8 percent. Consequently, fewer points will be made available in the next year. The average COP levels will have to be lower if the expected number of runs made in that year does not change. However, in that case the number of runs would again exceed the 1991 level by more than 8 percent, triggering another decrease in available points. If this sequence of reductions is taken into account, the total number of points available in 2003 is about 4 million instead of 7.7 million (excluding the ecopoints that are kept in reserve). Column 5 in Table 7.3 shows the COP levels required when these reductions are included in the calculations.

Table 7.3: Guestimate of expected and required average COP levels (in g NO_x/kWh) in Member States.

(1) country	(2) number of runs in 2003	(3) yearly growth	(4) required average COP level	(5) required average COP level special rule	(6) expected average COP level	(7) expected average COP level, trucks since '93
B	53,272	4.2	3.71	1.92	6.18	5.91
DK	60,811	3.5	4.05	2.09	6.18	5.89
FRG	747,202	3.8	3.92	2.03	6.21	5.93
E	2,494	6.3	2.92	1.51	8.54	6.18
F	8,683	4.8	3.50	1.81	6.15	5.89
GR	124,286	6.2	2.96	1.53	10.57	6.23
IRE	1,820	5.1	3.34	1.73	6.13	5.87
I	1,055,007	6.3	2.94	1.52	7.61	6.14
L	7,158	3.0	4.25	2.19	6.24	5.97
NL	207,022	4.4	3.63	1.87	5.53	5.47
P	819	6.2	2.97	1.53	10.19	6.23
UK	13,241	3.8	3.90	2.02	6.14	5.96
Total	2,281,815	5.0	3.38	1.81		

It remains to be determined what the actual average emission level of the truck fleets of the various Member States will be. This can be calculated if the lifetime function of the trucks in the various Member States and the COP level of the trucks acquired in each year are known. For the COP levels, it is assumed that trucks sold in a certain year conform to the European rules set out for that year. Consequently, prior to 1988 the COP level is 15.8 g NO_x/kWh. From 1988-1992, it is 14.4, from 1992-2000, 8.0 and from 2000, 4.0 (it is assumed that the so-called Euro 3 standard comes into force in 2000).

The lifetime function used to determine how many trucks are left in a specific year is the Weibull distribution,[6] estimated for each Member State by Samaras and Zierock (1992). For simplicity, it is assumed that the number of new trucks bought each year is a fixed percentage of the stock of the foregoing year.[7]

Column 6 in Table 7.3 shows the estimated average COP level of the truck fleet of each Member State to be expected in the year 2003. However, this average level need not represent the specific trucks used for the trans-Austria trips. In fact, it might be assumed that cleaner trucks are used. Column 7 shows the expected average COP level, assuming that only trucks are used in 2003 which where bought after 1993. In that case, the difference between the required level and the expected level is less, primarily because the high COP levels expected in some countries have been reduced.

These estimates give an indication of the potential for trade. First, it seems likely that in all Member States ecopoints will become a scarce commodity, implying that no opportunities would remain for reallocation of surplus ecopoints from some Member States to others via the European Commission. Second, assuming that the costs of reducing truck emissions rise with the gap between actual and required COP levels, trade will become attractive. Firms from countries where COP levels need to be lowered substantially, will buy ecopoints in countries with lower reduction necessities. So, the difference between the actual and the required emission levels may be taken as an indication of the required effort per country per truck. The greatest differences seem to exist in the four Mediterranean countries, due to high expected growth rates of their truck fleets as well as high actual emission levels. Greece and Italy in particular will have to shoulder a relatively large part of the burden, given the volume of their transit traffic. In Spain and Portugal, the difference between the required and the actual COP levels is high as well, but their volume of transit through Austria is very small.

There are, however, other factors as well which will encourage trade between firms, regardless of whether they are based in the same or in different countries. When a new truck is bought to replace an old one, the number of ecopoints earned depends on two factors. First is the difference between the COP level of the old truck and the new one. If this difference is large, the truck owner saves a larger number of ecopoints than if the old truck were less dirty. Furthermore, the number of ecopoints saved will also depend on the number of runs made with that particular truck. Each trip, a number of points is saved equal to the difference in COP levels. The more runs made, the higher the total amount of ecopoints saved. The value of saved (and marketable) ecopoints can be balanced against the costs of (eventually accelerated) replacement of old trucks. Conversely, firms may also choose to acquire ecopoints on the market, instead of having to buy new cleaner trucks before the old ones are written off. In this way, tradeability of ecopoints allows firms more flexibility. As already noted in the preceding section, incentives to trade will also emerge between firms having different growth rates.

Conclusions

The transit agreement between Austria and the EU, recently transposed into an internal EU regulation, is an innovative approach to reducing the environmental damage caused by traffic. As far as the reduction of NO_x emissions from trucks in the Alpine region is concerned, it will be an effective instrument. However, it is not an efficient approach, due to the lack of opportunities to equalize the relative scarcity of ecopoints between countries and between firms. When ecopoints become scarce in most, or even all Member States, the reallocation mechanism of surpluses between Member States will fail. Also, the administrative procedure to allot ecopoints to transport firms does not guarantee equalization of marginal abatement costs.

The agreement could be improved if the ecopoints were to be made tradeable. Abatement costs would then be minimized and firms would have a clear price incentive to reduce their NO_x emissions. It was not possible to calculate the potential cost savings. Nevertheless it seems plausible, given the different needs to reduce average emission levels, that considerable scope for trade exists. Allowing for trade could bring in substantial efficiency gains, as a result of optimizing the level of emission reduction between countries. Apart from this, an important benefit from trade would be that it allows firms more flexibility in selecting and utilizing least cost options.

Notes

[1] About 50 percent of the NO_x emissions in OECD Europe are caused by traffic (OECD, 1992). Another major contributor to acid rain are SO_2 emissions, which are mainly produced by industry and electricity generating companies.

[2] For example, under this type of rule electricity generating plants are obliged to limit emissions of pollutants to a certain amount per unit of electricity generated.

[3] Based on reports in issues of *Europe,* 8 February through 3 March, 1994.

[4] Austria, Finland and Sweden are not included in the calculations because the Samaras and Zierock study does not provide data for these countries. Consequently, no data are available on the development of the truck fleets in these countries and on their age composition.

[5] Using standard regression analysis, the following function has been estimated for the period 1978-1992: ln(transito) = a + b ln(EU GDP) + c D1. In this function, D1 is a dummy variable which represents the combined effect of the ban on night driving by trucks which was introduced in Austria at the end of 1989, the closure of the Kufstein motor bridge and long lasting strikes by Italian truck drivers and customs officers in 1990 and 1991. The year 1978 was chosen as a starting point because there was a change in trend that year. The result of this estimation is (T-statistics in parentheses):

$$\ln(\text{transito}) = \underset{[-1.95]}{-4.06} + \underset{[6.68]}{1.71}\ \ln(\text{EU GDP}) - \underset{[-3.97]}{0.19}\ D1$$

$R^2 = 0.876$
Durbin-Watson statistic = 1.56

[6] The Weibull distribution is an empirically established distribution of the service lives of technical products. See Samaras and Zierock, 1992, p. 282.

[7] In reality, new purchases of trucks fluctuate considerably: in times of recession, purchases usually fall, and pick up when economies are booming. However, taking this into account would seriously complicate the calculations.

References

Advies Europees Parlement over Verordening (EEG) nr. 3637/92 van de Raad, DOC 343(92).

CEC (Commission of the European Communities) (1992), *Publikatieblad van de Europese Gemeenschappen,* 21 December 1992, Nr. L 373/1 - 373/45 Verordening (EEG) nr. 3637/92 van de Raad.

CEC (Commission of the European Communities) (1993), *Publikatieblad van de Europese Gemeenschappen,* 25 February 1993, Nr. L 47/28 - 47/53, Administratieve Regeling.

CEC (Commission of the European Communities) (1994), *Publikatieblad van de Europese Gemeenschappen,* December 1994, Nr. L 341/20 - 36 Verordening (van de Commissie) nr. 3298/94 van de Raad.

Dwyer, J.P. (1991), 'California's Tradeable Emissions Policy and Its Application to the Control of Greenhouse Gases'. In: *Climate Change, Designing a Tradeable Permit System,* OECD, Paris.

Europe (1994), February 8, 10, 18, 28 and March 1, 2, 3.

Eurostat (1993), *Aussenhandel, Statistisches Jahrbuch,* Luxembourg.

Eurostat (1993), *National Accounts 1970-1991*, Luxembourg.

Hahn, R.W. and G.L. Hester (1989), 'Where Did All the Markets Go? An Analysis of EPA's Emissions Trading Program', *Yale Journal of Regulation*, Vol. 6, nr. 109, pp. 109-153.

'Haken en Ogen in de Transito-verdragen' (Problems with the transit treaty), *Europa van Morgen*, Vol. 22, nr. 34, 26 November 1992.

NEA (1992), *The Transport of Goods by Road and Its Environment in the Europe of Tomorrow*, NEA, Rijswijk.

OECD (1992), *Market and Government Failures in Environmental Management: The Case of Transport*, OECD, Paris.

Samaras, Z.C., and K.-H. Zierock (1992), *Forecasts of Emissions from Road Traffic in the European Communities*, Report for DG XI - Environment, Nuclear Safety and Civil Protection, Luxembourg.

United Nations (1971-1990), *Annual Bulletin of Transport Statistics for Europe*, Economic Commission for Europe, New York.

8 Prospects for European Environmental Tax Reform

Ruud A. de Mooij
Herman R.J. Vollebergh

Introduction[1]

In previous chapters of this book specific proposals for economic instruments in environmental policy, such as environmental taxes and tradeable permits, have been suggested and discussed. All of these proposals are directed towards achieving environmental targets set by the EU itself and by its Member States. Moreover, they also fit into calls for green taxes which sometimes appear in debates about environmental regulation and tax policy. Interestingly though, these claims usually do not ask for just a rise in environmental taxes but, rather, a *shift* in the overall tax burden of society. By giving an inevitable rise in the proceeds for the government, environmental taxes offer opportunities for reductions in other taxes. If other taxes - such as income taxes - are cut back with the revenues raised by the environmental tax, this shift is said to be 'neutral' with respect to the government budget.[2] We call these ideas of shifting from direct taxes on income towards taxes on energy and materials an exercise in *environmental tax reform*.

This chapter discusses prospects for such a European environmental tax reform taking account of its public finance implications. We do not distinguish explicitly between environmental taxes or auctioned tradeable permits as both these instruments raise public revenue which might be used for reductions in other taxes. Like the authors in the previous chapters of this book, we start from basic principles by assuming that the primary goal of an environmental tax reform is to internalize environmental externalities. In particular, environmental taxes or tradeable permits may improve the allocation of productive activities and consumption if these activities contribute negatively to the economic welfare of society through their effect on the environment. As current tax systems in the EU impose rather low taxes on those activities and environmental targets require serious additional policy measures in most countries of the EU, the use of environmental taxes has the potential to reap important welfare gains.

To illustrate, several environmental problems threaten Western European countries and sometimes the world as a whole - some of which are described earlier in this book such as climate change, acid rain and nutrient leaching. Environmental tax reform offers an opportunity to tackle those problems, thus striving to achieve environmental targets set by the EU or by its Member States. Calculations by the OECD, for instance, suggest that the Toronto agreement on climate change - viz. 20% reduction of CO_2 emissions by 2010 compared to the 1990 level - would ask for a European carbon tax of $213 per ton of carbon. This amounts to a rise in the price of oil and gas by 68%, 299% for coal, and 36% for electricity (Burniaux et al., 1991, p.29). If Member States were to take such a proposal seriously some of the current tax systems would be turned upside down. For instance, Greece would have to raise 21% of its total tax revenue from energy taxes compared to 8% currently. On average, the Union as a whole would then generate 11% of its total tax revenue from energy taxes compared to the current 4%.[3] But even with less ambitious plans, opportunities for a shift towards environmental taxes exist especially if environmental taxes or auctioned tradeable permits different from energy or CO_2 are considered as well - for instance the set of instruments proposed in this book.

Recent EU policy documents - such as Delors' last White Paper - also advocate the use of taxation in order to address certain environmental problems. However, the role that the EU has to play in this respect is often strongly disputed. In fact, this is a struggle over the interpretation of the subsidiarity principle as codified in the Maastricht Treaty. Some people are very reluctant to hand over control to the EU on environmental issues. Others, in contrast, advocate a strong role for the EU in order to coordinate or even harmonize tax policy in general and environmental tax policy in particular. They argue that environmental issues often transcend country borders - either in a physical sense or as packaged in traded products - so that coordinated action may be favorable for all the countries involved. This implies that there is a key task for the EU to play in tackling (inter)national environmental problems.

In evaluating prospects for an EU environmental tax reform properly, one should not only take environmental benefits into account but also its costs. Overall welfare effects of such a policy depend also on how employment, national income, and its distribution are affected. In particular, we discuss to what extent environmental tax reform might not only improve the environment but also yield positive or negative side-effects for the economy. These side-effects are discussed using the literature on the 'double dividend' hypothesis. This hypothesis suggests that an environmental tax reform would yield benefits to society over and above a cleaner environment, e.g. by alleviating current distortions in the tax system. Studies on the double dividend usually employ general equilibrium analysis and show how all kinds of interactions between markets, which are relevant in the context of environmental tax reform, influence the expected outcomes of the reform as envisaged.

The next section begins with a description of current tax profiles of Member States in order to give an impression of the potential for swaps of environmental taxes for ordinary income taxes in the European Union. Some statistical background information is provided on revenue-raising aspects of tax systems which is an unfortunate omission in the current debate. We classify tax profiles of the European Union as envisaged in the ideas of environmental tax reform and investigate the role environmental taxes currently play in European tax systems. Furthermore, current taxes are confronted with the environmental targets countries have agreed upon. Next, this reform is put into its tax policy perspective and we discuss how a tax shift relates to general tax policy principles known from public finance literature. Subsequently, possible indirect consequences of such an environmental tax reform in the EU are discussed. By using general equilibrium tax incidence analysis, we describe consequences for the environment, national production and income, employment and the distribution of incomes. In this way, threats and opportunities for an environmental tax reform are considered for the EU. Finally, the last section evaluates the prospects for a green tax shift, given the different proposals sketched in previous chapters of this book. The focus of this section is on the costs and benefits of environmental tax reform, the role the EU could play, and how the prospects for such a shift can be improved.

Developments

The analysis of tax profile developments focuses on environmental tax reform as a shift from income taxes to all kinds of environmental taxes or auctioned tradeable permits. In order to have some insight into the order of magnitude of such a reform, it is illuminating to have some knowledge of the current revenue-raising profiles of the countries in the EU. From the environmental perspective some insight into currently existing environmental taxes provides important information to what extent environmental problems are already taken into account by the tax system. These results can be set against the environmental targets revealed by the EU and its Member States.[4] If a gap between environmental policy goals and current environmental policy performance in Member States exists, current tax systems leave room for potential welfare improving tax shifts (see Chapter 1 for a further discussion of this important point of departure).

Developments in tax profiles of countries are usually presented from the revenue-raising perspective. We follow this practice using the OECD/IMF classification scheme (Messere, 1993, pp. 44ff) and the data set presented in the OECD/IMF Revenue Statistics which presents revenues raised by central governments. In particular, the tax base is split into several categories: income and profits, payroll, property, goods and services, and others. More detail is obtained by examining who exactly pays the tax, i.e. households or incorporated enterprises. Compulsory payments to governments, earmarked for social security expenditures, are the only exception to both principles and are presented separately. Besides, payroll taxes and only play a very modest role in overall tax revenue (less than 1%) and are therefore omitted in Table 8.1. With respect to the countries on which data are presented we also include the three youngest Member States, Sweden, Finland and Austria. Developments in the different sources of the general tax base are summarized in Table 8.1 and Figure 8.1.

Table 8.1: Tax share developments EU 1965-1991 (unweighted average of receipts).

	1965	1975	1985	1991
Taxes on income, profits, capital gains (1000)	28	33	33	34
- individuals (1100)	22	27	27	28
- corporations (1200)	6	6	6	6
Social security contributions (2000)	23	28	28	28
Taxes on property (4000)	7	5	4	4
Taxes on Goods and Services (5000)	36	30	30	30
- general taxes (5110)	13	15	17	18
- taxes on specific goods and services (5120)	23	15	13	12
Other taxes	6	4	5	4
Total taxes	100	100	100	100
in % of GDP	28.4	34.8	40.7	42.0

Source: IMF Revenue Statistics

Table 8.1 reveals that overall tax receipts rose between 1965 and 1991 from 28.4% to 42% of GDP. In particular, the share of personal income taxes and social security contributions grew considerably. Its share in terms of total tax receipts grew between 1965 and 1975 from 45% to 56% while it remained almost constant thereafter. The share of so-called general consumption taxes, being the Value Added Tax and sales taxes, grew at an almost constant rate from 13% to 18% of total tax receipts at the expense of taxes on property (and wealth) and specific taxes, such as excises on tobacco and sugar. The share of total taxes on goods and services, however, declined from 36% to 30% in this period, although it remained constant on average, since 1975! Figure 8.1 reveals how striking these patterns are. In light of the current debate on environmental tax reform, the trend of tax profiles in the last decade away from specific taxes towards more general consumption taxes implies that environmental tax reform is a plea for a reversal of an ongoing trend.

Table 8.2 offers some interesting information about taxes which are to be found in the category of taxes on goods and services, among them energy taxes or taxes on hydrofuels. One can derive from the table that the decline in the share of taxes on specific goods and services in total tax revenue is explained by the decline in specific excises as well as the fall in customs and import duties. The latter development is not surprising as the European Union had a central aim to abolish these duties. However, the decline in the relative share of specific excises is remarkable. This decline is mainly explained by a decline in revenues generated by excises on tobacco and alcohol, but hydrocarbons, on average, also raise less revenue as a percentage of GDP in most Member States (Messere, 1993, p. 420). Moreover, recurrent taxes on motor vehicles have been rather constant in the same period.

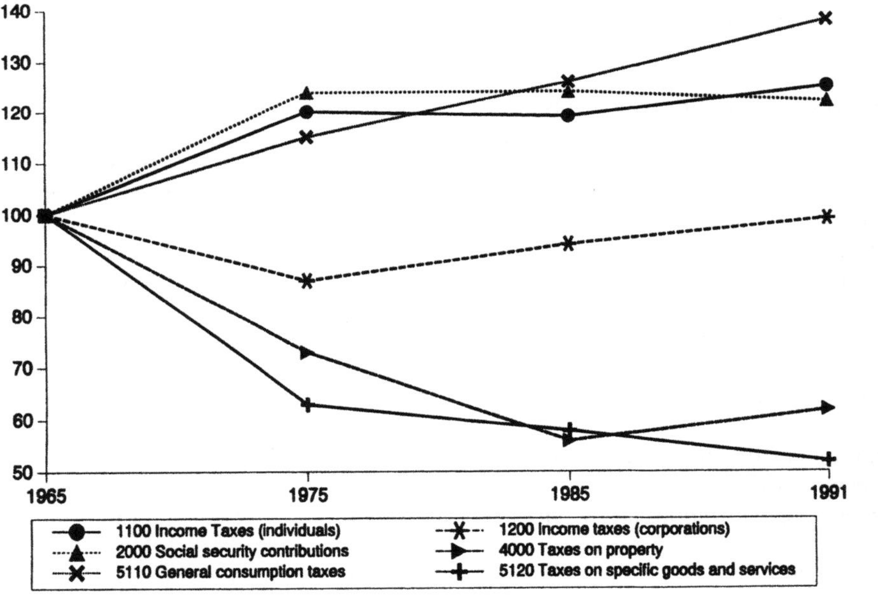

Figure 8.1: Tax share developments EU (unweighted average of receipts; 1965 = 100) (IMF Revenue Statistics).

Table 8.2: Developments in taxes on specific goods and services in the EU, 1965-1991.

	1965	1975	1985	1991
General taxes (5110)	36	50	54	58
Excises (5121)	39	32	30	28
Customs and imports duties (5123)	14	7	4	3
Other specific excises (5122 + 5124..8)	6	6	7	6
Recurrent taxes on motoring (5211..12)	3	3	3	3
Other specific taxes (5213+5300)	2	2	2	2
Total taxes on goods and services (5000)	100	100	100	100
in % of GDP	10.6	10.7	12.7	13.1

Source: IMF Revenue Statistics.

These stylized facts already hint at the fact that, from the revenue perspective, certain specific taxes which could be considered as environmental taxes - such as taxes on hydrofuels ('energy taxes') or taxes on motor vehicles - are unimportant in the EU. This conclusion is even more profound if we look at Figure 8.2. This figure presents a cross section of hydrofuel taxes, recurrent taxes on motor vehicles and other explicit environmental taxes as percentages of either taxes on goods and services or total tax revenues in the different Member States of the EU for the year 1991. Indeed, taxes on hydrofuels and motor vehicles only raise, on average, 5% of total tax receipts in the EU, rendering them unimportant as revenue-raising devices. Even within the category of taxes on goods and services (5000) the share of taxes on hydrofuels is, on average, only 12.8%, with a maximum of 24% in Italy. Other environmental taxes play almost no role as a revenue-raising instrument. Nevertheless, the figure shows remarkable differences between Member States with respect to the use of specific taxes with an environmental dimension. The receipts raised by the tax on hydrofuels as a percentage of total taxes range from 0.3% in Luxembourg to almost 9% in Portugal. This, at least in part, reflects the slow progress in the harmonization process of tax rates on hydrofuels. Moreover, it seems that southern Member States of the EU depend relatively more on such taxes than northern Member States.

The preceding analysis only provides information on a very general level, for no insight is gained into either the importance of the overall tax burden in the individual Member States (some countries like the Netherlands and Sweden face relatively high tax burdens, while others like the United Kingdom and Spain do not), or into the interplay of different taxes such as specific taxes on goods and the Value Added Tax, which both raise end-use prices of goods. Therefore, further information about the importance of taxes for the environment can be derived from Table 8.3. This table presents the total amount of taxes on energy products (as the addition of different excises and VAT or sales taxes) as a percentage of gross product prices for either industry or consumers. The differences in tax treatment of energy products between countries is remarkable as these figures show. Gasoline taxes differ considerably, but also taxes on other energy products such as electricity or natural gas show a large variance between countries. Besides, industry is treated rather differently from consumers in all countries.

The general picture that appears is that specific environmental taxes and other revenue-raising environmental instruments play only a very modest role as revenue raisers for central governments nowadays. Even if one takes a broader view on environmental taxes,

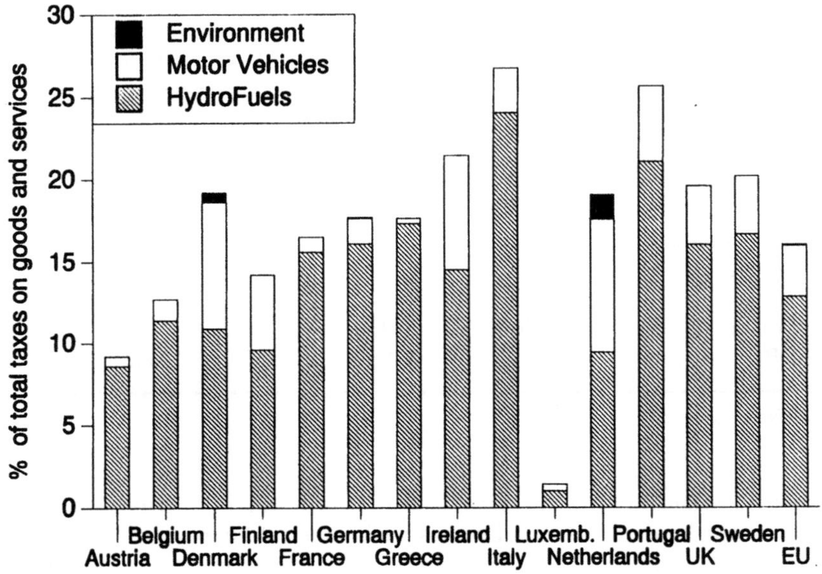

Figure 8.2: Tax share for some specific excises in different EU Member States, 1991.

including taxes on hydrofuels and motoring, the relative importance of these taxes in total tax revenue seems far from impressive. The picture, however, changes in a fundamental way if one allows for the absolute amount of money raised by these taxes and the existence of VAT raising all prices of consumption goods by a similar percentage. As can be derived from Table 8.3, specific excises and VAT together raise the net price of gasoline by about 233%! However, other taxes on several energy products are much lower.

Further reflection reveals that our figures do not tell the whole story.[5] For instance, the environmental tax base should ideally also include other taxes and tax exemptions such as exemptions for commuting in income taxes. Another pressing problem is that countries have their own classification structure with respect to various means of government intervention, especially with respect to tax expenditures, taxes paid by the government itself and activities classified as provided by the (central) government. Local taxes are also excluded, while, in practice, several environmental taxes are levied by local authorities. Furthermore, (government) prices of public utilities are not included though they usually provide interesting opportunities for environmental tax reform as well. Therefore this analysis can only be considered a first step. More reliable data on 'environmental taxes' not recognized by the data set we used, should also be included, such as revealed by recent surveys of the OECD (1993 and 1994) and IEA (1994).

Thus, from the environmental point of view, our ratios should be seen as a preliminary rough estimate. Nevertheless, our analysis has the advantage over earlier studies of including *implicit* environmental taxes such as currently existing excises on hydrocarbon oils. These taxes already internalize part of the environmental problems in the status quo. Previous surveys of economic instruments employed a more narrow approach (OECD, 1989). In particular, taxes were classified as 'environmental' if and only if the legislator had expressed explicitly that this tax serves some environmental purpose. Fortunately, the more recent surveys mentioned above are less strict in their definition of 'environmental taxes' and, therefore, more appropriate from an economic point of view.

Table 8.3: Amount of tax in prices of energy products in 1991(in percent).

	Industry			*Consumers*			
	natural gas	heavy fuel oil	light fuel oil	coal	natural gas	gasoline	electricity
Austria	0	7	21	0	17	65	17
Belgium	0	0	0	0	15	72	16
Germany	11	13	15	0	17	74	19
Denmark	n.a.	0	0	0	18	66	48
Spain	n.a.	12	27	n.a.	n.a.	69	11
Finland	2	7	4	8	19	71	18
France	0	21	34	0	13	78	20
UK	0	13	10	0	0	68	0
Greece	n.a.	41	42	n.a.	n.a.	75	15
Ireland	0	9	19	n.a.	11	66	11
Italy	9	39	66	n.a.	41	74	17
Luxembourg	n.a.	2	n.a.	n.a.	6	66	6
Netherlands	4	17	23	0	17	73	16
Portugal	n.a.	33	56	n.a.	n.a.	74	15
Sweden	n.a.	64	48	45	n.a.	75	32

Source: OECD Energy Prices and Taxes.

Principles

This section introduces some tax policy principles relevant in the context of environmental tax reform.[6] Here we restrict ourselves to the analysis of taxes as revenue-raising instruments, although, as noted before, auctioned tradeable permits could be included as well. First, we introduce the fundamental characteristics of any tax, namely the revenue-raising effect and the regulatory effect. Subsequently, we discuss tax principles related to both effects. These include the benefit principle and the ability-to-pay principle that mainly relate to the revenue-raising effect. The tax neutrality and corrective taxation principles have a closer link to the regulatory effect. A final principle is the simplicity of taxation, which is necessary in order to have effective taxes in general.

Tax Consequences

To keep track of the effects of taxation on the choices of economic agents, economists tend to compare price-dependent production possibilities with and without some kind of tax. The consequences of taxation can generally be separated into a revenue effect and a regulatory effect. The revenue effect amounts to the fact that taxes usually have the function of financing government expenditure. In particular, taxes raise public revenues, thereby implying a transfer of money from tax liable agents to the government. The government, in turn, uses these tax revenues by providing agents with public goods and services. This provision of public goods and services is called the productive effect of taxation.[7]

At the same time, however, any tax regulates in some way or another through its impact on relative prices. In particular, the regulatory effect of taxation can be viewed as the interplay of two effects central in economic analysis, namely, the substitution effect and the income effect. The substitution effect denotes a shift in behavior away from taxed commodities. If, for instance, taxes on leaded gasoline are imposed while unleaded gasoline is not taxed, the price of leaded gasoline rises relative to unleaded gasoline. As a result, agents are expected to shift from cars driven on leaded gasoline to cars driven on unleaded gasoline. The income effect represents the loss in 'purchasing power' as a consequence of the tax. In our example, the income effect exerts a decline in the demand for both leaded and unleaded gasoline provided these goods are normal goods.

It is important to note that both the revenue effect and the regulatory effect occur with *any tax*. This fundamental observation also applies in the context of environmental tax reform. As discussed in the previous section, it is sometimes thought that a tax can be called an environmental tax if and only if the legislator has expressed explicitly that this tax should serve some environmental purpose. In contrast, others seem to think that taxes meant for revenue-raising purposes would not regulate. Both statements are false. All existing taxes raise revenues and change behavior, thereby regulating the environment either implicitly or explicitly. Any policy analysis has to acknowledge this essential point of departure.

Tax Principles

Tax policy principles relate directly to the revenue and regulatory effects of taxes. In particular, the *benefit principle* and the *ability-to-pay principle*, relate to the revenue effect. First, the benefit principle is shorthand for the view that taxes should be levied in accordance with the benefits arising from the government services financed by the tax. In other words, if someone collects the productive effect, she has to pay for it. Second, the ability-to-pay concept states that the highest taxes should be levied on those who are most able to bear the cost of government spending, usually measured by income or wealth. Two other principles, tax neutrality and corrective taxation, relate to the regulatory effect of taxes. In our context of environmental tax reform, these tax policy principles are more important than the previous ones.

Tax neutrality expresses the view that taxation should impose the smallest possible impact on private decisions. These behavioral impacts of taxes are called distortions because taxes distort price signals in their role of reflecting relative scarcity of factors and commodities. Accordingly, these taxes are called distortionary taxes. They change the allocation of production and consumption as choices of private agents are based on distorted prices. For instance, if households start buying unleaded gasoline instead of leaded gasoline because the government imposes an additional environmental tax on the latter fuel, agents may buy another car which is adapted to unleaded fuel. From an economic point of view, such a change in the allocation of factors and commodities amounts to a shift away from the efficient allocation, i.e. the allocation in a non-distorted world. In this way, distortionary taxes reduce economic welfare and impose an additional burden on private agents. Intuitively, this burden amounts to the costs that private agents make to avoid taxes. Contrary to the costs for private agents of paying taxes, the costs of avoiding taxes are not associated with tax revenue. Consequently, these costs are usually called the excess burden of taxation, i.e. the burden that is not associated with any tax revenue for the government and thus with any public spending. Hence, the excess burden involves additional costs for society as a whole as, in contrast to the revenue-raising effect of the tax, they do not allow for the provision of public goods.

There is one type of tax that is not distortionary by definition. This so-called lump-sum tax does not impose an excess burden on private agents. A tax system based on lump-sum taxes is neutral by definition as well.[8] The literature on optimal taxation has pushed the tax neutrality principle to its extreme by taking the minimization of the excess burden with respect to tax rates as its ultimate goal - thereby abstracting from lump-sum taxes as this would yield a trivial outcome. It is in this context, that the Ramsey approach to taxation might be placed. This approach states that - relative to some social welfare function - the excess burden is minimized if taxes are levied in accordance to the lowest possible regulatory effect (Atkinson and Stiglitz, 1980). In a tax system based on indirect taxes only with zero cross-price effects, the familiar Ramsey rule boils down to the principle that it is optimal to tax commodities inversely proportional to their elasticities of demand. In other words, commodities with the lowest price elasticities should be taxed most heavily.

Contrary to tax neutrality, *corrective taxation* seeks to optimize allocative efficiency of society through regulation by taxation. To illustrate, if certain economic activities cause environmental damage, they are said to cause negative external effects. As far as these adverse effects on the environment are not taken into account by private agents, the market fails in its role to generate prices which reflect relative scarcities in a correct way. By changing price signals, taxes may induce economic agents to change their behavior in a more environmentally friendly way. These so-called Pigovian taxes are thus said to internalize the negative externalities associated with certain economic activities, thereby improving welfare. Note that the reasoning with respect to optimality of corrective taxes may be reversed compared to the tax neutrality principle. Although environmental taxes might also cause a rise in the excess burden of taxation (i.e. the *private* costs that private agents make to avoid paying taxes), they are first of all beneficial by producing a cleaner environment (i.e. the benefits that private agents experience due to the services from the environment as a *public* good). Hence, if these social benefits exceed the costs, economic welfare of society improves rather than declines due to the regulatory effect of the tax. In our example, the tax on leaded gasoline is said to internalize the environmental externality associated with leaded gasoline consumption in the behavior of agents. Indeed, as consumers shift their behavior away from the consumption of leaded gasoline towards unleaded gasoline, they may enhance rather than reduce economic welfare through the overall effect on environmental quality.

A final important tax policy principle relates to the *simplicity* of taxation. In order to be effective and enforceable, taxes should be simple and easy to monitor. In worlds where information is not costless and complete, trade-offs might exist with respect to theoretical optimality and the transaction costs related to taxes. This is not only true for ordinary income taxes or the Value Added Tax (VAT), but also for environmental taxes.[9] As shown in the previous section, there has been a growing trend in the countries of the EU, to shift taxes more and more toward those that are based on transactions and away from those that are not (Kay, 1990, p. 30). VAT and social security taxes fall into the former category, whereas taxes on wealth and capital income fall into the latter. The striking reliance of modern tax systems on VAT is to be explained by its relatively easy enforcement aspects: the tax base is usually clear, transaction-oriented and comprehensive.

With respect to this simplicity principle it is important to also take costs of implementation, administration and monitoring into account (see also Chapter 1). Like other instruments for environmental policy, taxation and permit regulation impose such costs as well. In order to attain a reliable comparison of instruments, any regulatory scheme should take other forms of regulation and their associated transaction costs into account. Thus, a particular tax policy cannot be rejected simply on the basis of its associated transaction costs, as the overall principle for evaluating different forms of regulation - taxes or other instruments - is simply which policy passes any overall welfare test, being the Pareto principle or the Kaldor-Hicks compensation test. Moreover, the choice of the tax base is also influenced by the magnitude of the transaction costs and its associated benefits (Vollebergh, 1994). Whether a specific environmental tax reform would be a welfare-improving exercise in practice is therefore not easy to answer.

Consequences

How are the consequences of an environmental tax reform to be assessed? This section discusses the complications arising from acknowledging the fact that environmental taxes do two things. First, they regulate behavior at the margin thus improving the quality of the environment. This is called the first dividend. Second, they generate revenue. The proceeds can be used in several ways, e.g. through producing public goods (education, public investment) but also by lowering other distortionary taxes (revenue recycling). This might allow for a second dividend by the alleviation of other pre-existing inefficiencies in the economy. To what extent both dividends are obtained is an issue taken into consideration by economists only recently.[10] We start by presenting the two dividends. Subsequently we discuss the double dividend issue of environmental tax reform using the results from recent analyses that employ general equilibrium analysis of tax incidence. The final subsection discusses threats and opportunities of environmental tax reform, given the general framework and by discussing some empirical findings.

Dividends and Double Dividends

Environmental taxes aim at correcting for environmental externalities. In theory, these taxes are able to achieve a given environmental objective in the most cost-efficient way. The condition for this theoretical result is a first-best world without any distortions different from the environmental externality. Moreover, as Chapter 1 showed, such taxes might be Pigovian provided that the regulatory authority knows the benefit schemes in the optimal state and that this is reflected in the tax (Baumol and Oates, 1988). The internalization of the environmental externality implies a welfare improvement and might be called the first dividend. This dividend is usually conditional on an analysis where it is assumed either explicitly or implicitly that there are no welfare consequences of the revenue raised by such a tax. In other words, this analysis concentrates only on the welfare gains associated with the regulatory effect of a corrective tax. Theoretically this means that it is implicitly assumed that the proceeds of the tax are distributed in a non-distortionary lump-sum fashion across the population at large. Furthermore, no attention is paid to other distortions in the status quo, whether they are due to other externalities, distortionary taxes, imperfect competition, informational restrictions, or political economy matters.

The limitations have not prevented some politicians and economists from trying to generalize its predominantly partial results. They suggest that, in a world where all kinds of pre-existing market distortions exist, environmental taxes would yield welfare gains *over and above* the positive welfare effects of a cleaner environment. In particular, they argue that, if the proceeds of the environmental tax are used to reduce other distortionary taxes - such as labor taxes - this would yield another dividend, for instance, the alleviation of labor market distortions. Thus, besides the first dividend, a second one could be obtained if the revenues from the environmental tax are used to reduce pre-existing distortionary taxes. This is essentially the *double dividend* argument which is also prominent in Delors' White Paper mentioned in the introduction: an environmental tax reform would not only improve the environment but also raise employment through lower labor costs (net of tax) and therefore reduce unemployment. If this double dividend argument held, an environmental tax reform would be very attractive for policy makers. They might even no longer need to bother about the value of the environmental benefits - which are rather difficult to measure - because environmental taxes would already be favorable from a non-environmental point of view. This view is reflected in particular by Pearce (1991) and Oates (1991) who have argued that, in the presence of distortionary taxes on labor -

which cause high excess burdens - environmental taxes may even be more attractive instruments than in the first-best world where they only guarantee the first dividend.

Several authors have raised doubts about the *double* dividend just mentioned. They argue that the partial equilibrium analysis - which explores the environmental distortion and the labor-market distortion separately - might be rather misleading. In particular, in the case of the double dividend, the partial equilibrium approach ignores the interaction between environmental taxes and their effect on the initial labor market distortion. This interaction can be analyzed by combining second-best analysis with a general equilibrium analysis of tax incidence. Following the seminal paper of Harberger (1962), tax incidence analyzes the question 'who bears the economic burden of taxes', for agents who *pay* taxes are not always the same as agents who *bear* the economic burden of taxes. For instance, if a tax is imposed on pollution, the burden of this tax is not necessarily borne by the supplier of the polluting input but by the polluter who is also an employer, an employee, a capital owner or a tranfer recipient. In other words, the factor on which the tax is imposed (environment) is not always the same as the household or firm who bears the economic burden of the tax in terms of lower real after-tax incomes. What exactly happens to the economy due to burden shifting depends on how economic agents react. Such reactions depend on conditions of demand and supply in the markets taken account of, the structure of these markets, expectations of agents and the time period for adjustments to occur. Thus general equilibrium analysis is able to investigate the ultimate economic effects of burden shifting, which denotes the process according to which agents respond to tax changes in different markets.

Using such a model, Bovenberg and De Mooij (1994a) show that such interactions can be crucial for conclusions about the double dividend (see also Bovenberg and Van der Ploeg, 1994, and Parry, 1994). They conclude that environmental taxes might even exacerbate, rather than alleviate, pre-existing labor tax distortions. Intuitively, the incidence of the environmental tax is not borne by the environment but by households. Accordingly, the people who ultimately bear the burden of the tax which is raised (the environmental tax) are the same people who benefit from the reduction in pre-existing taxes (the labor tax). An environmental tax reform thus amounts to a shift from direct taxes on labor income towards indirect taxes on the same labor income. For the consequences of a tax reform on employment this composition of the tax burden between direct and indirect taxes does not really matter - provided that there are no distributional changes due to the tax shift. What matters is the total level of the tax burden.

Given the constraint of an unchanged distribution of income between people, initial distortions may even be exacerbated by an environmental tax reform through its behavioral impact. The reason is that an environmental tax reform from labor to environmental taxation amounts to a shift from broad-based taxes to narrow-based specific taxes. This will change behavior by provoking tax evasion (which is the aim of the environmental tax), thus eroding the tax base and causing the tax system to become more distortionary as a revenue-raising device (i.e. from a non-environmental point of view). Intuitively, the change in behavior requires efforts from the regulated agents who try to avoid the specific tax. These efforts reflect the costs of a cleaner environment made by the regulated private sector. Accordingly, by increasing the excess burden on private agents, environmental taxes reduce the efficiency of the tax system as a revenue-raising device. Hence, a trade-off exists between the environmental aim and the distortions associated with raising public revenues. Whether the double dividend argument holds thus depends crucially on how agents react to the environmental tax reform. The Bovenberg and De Mooij model shows that there are circumstances in which the double dividend does not occur at all, a conclusion that is obtained by other models as well.

The possible failure of the double dividend does not mean that environmental taxes are inappropriate instruments for achieving environmental targets nor that the revenues from the environmental tax should not be used to cut other distortionary taxes. First, by employing taxes rather than other regulatory devices such as uniform regulation, the costs of achieving the environmental target might still be lower. As long as the benefits of reaching this target are positive and exceed the costs, environmental taxes are still a rational policy response. Furthermore, with respect to the recycling issue it is also preferable that the revenues of the environmental tax be used to cut distortionary taxes instead of returning the proceeds in a lump-sum fashion. The only correct conclusion from this general equilibrium analysis is that, in order to explore the consequences of environmental tax reforms, notice should be taken of the interactions between different market distortions. Besides, environmental improvements due to environmental tax reform may not be free but need to be weighed against certain costs. The above analysis on the double dividend argument is rather limited in the sense that it considers only the interaction between two specific distortions, namely, the environmental and the labor market distortion in a specific type of second-best world. In reality, all kinds of other distortions may be present as well and this may leave room for both opportunities and threats for the economy. These will be discussed below.

Threats and Opportunities

What consequences are to be expected for the EU if an environmental tax reform is envisaged as a serious policy option? This question is central in the growing number of studies based on tax incidence analysis, in which all kinds of possible mechanisms relevant in the case of environmental tax reform are uncovered.[11] The main points in these studies are the extent to which agents are able to shift their tax burden to others, whether existing tax systems are optimal or not and how mobile the production factors are. Because the EU is not a closed economy, the effects of environmental tax reform are largely dependent on what exactly happens to those production factors which are mobile internationally. The mobility of capital is of particular concern for Member States who intend to impose environmental taxes. Another issue of importance is the assumption of income neutrality, i.e. refraining from the effects on the income distribution.

Environmental taxes inevitably reduce the profitability of a number of firms in the EU. This may stimulate these firms to leave the EU and move to other countries without such environmental taxes. This process might even exacerbate global environmental problems if local conditions outside the EU are less strict (Hoel, 1992). Furthermore, with low profits, financial capital will flow abroad so that several activities in the EU might become less attractive thus stimulating activities in other parts of the world, e.g. in Southeast Asia or Eastern Europe. Higher prices due to environmental taxes may also induce higher export prices, thus shrinking competitiveness of the European export sectors on world markets. Finally, domestic markets face higher costs of pollution thus causing higher output prices. This may trigger unions to claim higher wages to compensate for this effect. Wage increases, in turn, may provoke producers to raise their prices in order to compensate for the higher costs of labor, thereby setting a wage-price spiral in motion. In short, environmental tax reform might reduce the competitive strength of currently polluting firms within the EU and this might have adverse effects on production and income in the EU through a number of channels.

We must be careful about jumping to conclusions here. First, at least some of the negative effects will be counterbalanced by the gain in relative competitiveness of the pollution-extensive industries which benefit from the tax shift. Higher taxes on pollution induce behavioral changes at the margin directed towards less polluted products. Moreover, the tax *shift* supports this positive effect as producers facing high labor costs in the status quo will benefit from lower labor taxes.[12] Furthermore, investments in cleanup activities and abatement may raise the demand for investment goods, thus stimulating economic activity. Such Keynesian demand effects may be favorable for national production and income, especially in the short run. Another opportunity is due to a first-mover advantage. If the EU imposes an environmental tax, firms will seek new, cleaner technologies. When other countries impose environmental taxes at a later date, the EU will have a comparative advantage in its cleaner technology and may reap the welfare gains associated with the fist-move at a later date. Finally, a cleaner environment not only yields higher environmental services as a consumption good but also increases the supply of the environment itself in its role as a production factor. For instance, by increasing the fertility of land or reducing illness of workers due to air pollution, a cleaner environment raises the productivity of land and labor, thus boosting economic activity.

Apart from these direct and indirect effects of the tax shift, one can also try to influence who exactly bears the tax burden. An attractive opportunity to yield beneficial effects from an EU perspective is to shift (part of) the tax burden to foreigners. To illustrate, if an energy tax is part of environmental tax reform, the overall reduction in energy use of EU member states may exert a negative impact on world energy prices. Accordingly, the economic burden of the energy tax in the EU can partly be shifted towards the suppliers of energy. Environmental taxes may also raise world output prices. Intuitively, they may reduce the supply of particular goods in EU economies, especially those that are energy-intensive. This may exert an upward effect on the world market prices for those commodities. Hence, importers of those commodities may bear part of the economic burden of the environmental tax in the EU. Whether burden shifting to outsiders is realistic depends ultimately on the expected behavioral responses of the suppliers of the taxed commodities.

Another opportunity is to shift the tax burden to an appropriate group of insiders, viz. the different polluting countries and income categories within the EU. Thus far, we have ignored the fact that the tax shift may involve different economic agents in a distinct manner. Indeed, tax shifting between different factors of production (from labor towards pollution) or between consumption commodities may sometimes also imply a shift in the incidence of the taxation between various income categories, e.g. between households within a given jurisdiction. Such a shift in the incidence of taxation might cause additional economic effects compared to the previous analysis which may be either favorable or unfavorable for the economy. In particular, if the initial tax system is inefficient from a non-environmental point of view, environmental taxes may be an indirect way to alleviate such inefficiencies. This argument will be illustrated with an example in which we assume that the environmental tax reform consists only of a rise in energy taxes and lower taxes on labor (Bovenberg and De Mooij, 1994b).

Consider a rise in the tax on household energy use and a simultaneous reduction of the tax on labor (e.g. by raising the tax credit for workers). On the one hand, the burden of the energy tax is distributed among people according to their use of energy. As consumption patterns might be rather different for retired people, disabled or unemployed workers, employed people, capital owners, etc. the burden of the energy tax is distributed unevenly among consumers. In particular, lower income groups (transfer recipients) and elderly people usually spend a larger share of their budget on energy than others. Hence, they may experience a relatively large burden of the energy tax.[13] On the other hand, if the government uses the revenues of the energy tax to compensate only one specific group of people (i.e. workers) for their loss in real income, while other people (i.e. those who depend on non-labor incomes) are not compensated at all, the environmental tax reform amounts to a shift in the overall burden of taxation from one group of people to another. In particular, transfer recipients face a loss in purchasing power while employees may even benefit from the environmental tax reform. In other words, the environmental tax reform amounts to an indirect way of redistributing incomes from transfer recipients to workers. Standard economic theory states that such a redistribution from inactive people to active workers may stimulate economic activity and, more specifically, employment. Hence, this would imply a double dividend (i.e. more employment and a cleaner environment). Note, however, that this double dividend is only achieved at the cost of a change in the income distribution in favor of workers, which is a consequence of the choice not to compensate polluters in accordance with their income levels. In this way, environmental tax reform can be interpreted as an indirect way to alleviate labor market distortions which are initially present for equity reasons.[14] This redistribution from the poor to the rich may be difficult to realize politically.[15]

Several empirical studies have also been used to estimate whether there is support that an environmental tax reform - from labor and capital to energy - yields a second dividend over and above the environmental dividend (see De Wit (1994) for an overview of empirical studies). Goulder (1995), for instance, uses a dynamic computable general equilibrium (CGE) model calibrated for the US economy in order to investigate the distortions associated with several taxes. He finds that environmental taxes are typically more distortionary than taxes on labor and capital. The OECD GREEN model for the world economy also reveals that an environmental tax reform in the EU would reduce employment and national income, although these effects are rather minimal. The econometric QUEST model of the European Commission showed initially that an environmental tax reform would have little impact on employment (EC, 1992). After a revision of the model, however, the effects are much more positive: an environmental tax reform in the EU would raise employment and national income by about 2% (Koopman, 1994). The Dutch Central Planning Bureau also computed the consequences of an environmental tax reform for the Dutch economy by employing a multisectoral econometric model (CPB, 1992). Their conclusions vary with the assumptions about the number of cooperating countries where the tax reform is implemented, who exactly is subject to the tax (firms and/or households) and how the revenues are recycled. If, for instance, an environmental tax reform in the Netherlands is envisaged, a positive effect on employment is obtained if the tax reform is limited to households only. With firms also being subject to the tax, the conclusions are reversed. Both a unilateral tax and an OECD-wide tax would reduce employment and incomes substantially. To conclude, not only theoretical but also empirical models indicate that both threats and opportunities exist for production, income and employment.

Prospects

The first and most pressing question is: what kind of environmental tax reform should be implemented at which policy level? We noted in the introduction that opinions strongly differ on this issue. The struggle focuses on how the subsidiarity principle, as it is now codified in the new treaty of the EU (Maastricht Treaty), is to be understood. Some observers think that subsidiarity might prevent further expansion of environmental policies at the EU level. Others, in contrast, believe that the Maastricht Treaty as a whole, might take the responsibility for environmental policy away from the Member States themselves. From the economic perspective it seems preferable to take policy measures at the level where costs and benefits balance (compare, for instance, Oates and Schwab, 1988). For example, a certain Member State would be more reluctant to reduce emissions if the benefits accrue not only to this country but also to its neighbors: why should a country act to improve the environment in other countries if the relation with local benefits is rather weak? As is well known from the theory on public goods, if all countries think and act this way, the total provision of public environmental goods may be too low which poses a serious threat for several environmental issues.

We will not go into the subsidiarity issue in detail here.[16] Only two relevant observations will be mentioned. First, the larger the scale on which environmental problems occur, the more relevant is the role for coordinated action, e.g. at the EU level. Indeed, several environmental problems cross borders in a physical sense (acid rain; climate change). By organizing coordinated action, the EU might prevent an underprovision of environmental goods. Note that coordination does not necessarily imply the use of policy instruments by the EU: coordination of local initiatives might work as well. Second, as long as countries compete on unequal terms of trade, policy intervention by the EU can also be beneficial. This is a double-edged sword. On the one hand, countries might pollute the environment in order to gain a competitive advantage. This may justify some intervention in order to protect those countries trying to internalize their externalities, for this might imply a loss of competitiveness if coordination is absent. On the other hand, countries might easily use environmental instruments in order to protect their own industries, which also justifies policy intervention by the EU.

The preceding analysis enables us to judge the prospects for environmental tax reform (both taxes and auctioned permits) in the EU, given the opportunities sketched in the previous chapters for such a reform from the environmental point of view. The core principle of an environmental tax reform is the *corrective* tax principle. The main reason for imposing environmental taxes should not be to raise revenue to finance government expenditures but to curb pollution and change undesirable behavior. The previous chapters have shown that, if the environmental goal is the leading principle in the design of such a reform, one might expect positive results for the environment.[17] Prospects, however, also depend on the non-environmental consequences of such a tax reform. Accordingly, these side-effects - which may imply threats and opportunities - should also be considered when evaluating the prospects for environmental taxes.

As turns out in the second section, environmental taxes take up only a relatively small part of the total tax burden currently. Nevertheless, in the light of the environmental targets that European countries have imposed on themselves and on the EU as a whole, environmental taxes are promising instruments for achieving these targets. Several opportunities exist at the level of the EU to implement policies for the benefit of the environment. Starting with energy and energy-related pollution, the EU could, for instance, implement several measures building on current energy tax trends. Indeed, although the exact magnitudes are unknown, changes in energy prices induce behavioral changes away from polluting activities as was clearly demonstrated as a result of the oil price shock in the seventies. Moreover, in the last few years, the opposite effect has prevailed due to the current low energy prices: emission and energy intensities are rising again. Hence, energy taxes may well raise economic welfare by internalizing the adverse external effects of energy consumption in energy prices. This is especially true because the initial taxes on energy are rather low which implies that they are currently far below the level which internalizes the environmental externalities.

As discussed in previous chapters, important opportunities exist for the EU to take the lead in this field. It could be a first step in the direction of an OECD-wide or even global green-house policy. Looking at the vast differences between countries - not only with respect to their energy tax rates, but especially with respect to differences in tax exemptions between products and industries (OECD, 1993) - the EU could broaden its scope for harmonization or coordination proposals. In particular, the rates of mineral oil excises have been subject to such a program, ignoring other differences between the excise systems of Member States (compare, for instance, Table 8.3). As a result the EU allows certain Member States to subsidize different industries incoherently, thus violating the first principles of the EU,[18] but also tax neutrality principles within a given jurisdiction.

The gradual allowance of environmental objectives in the program for harmonization of excise duties on hydrofuels is therefore certainly to be applauded, although it should be considered only a first step from the environmental perspective (Vollebergh, 1995). This allowance of environmental objectives might also be considered a model for other currently existing taxes, such as taxes on motor fuels and motor vehicles. Here, the EU could also enact harmonization proposals which take account of differences in pollution intensity of cars, something rather easy to implement because most countries already levy taxes on these objects. Furthermore, the EU could take the exhaustive taxation of energy and energy products seriously, including all fossil fuel products like hydrofuels, natural gas and coal. In particular, coal is currently untaxed or even subsidized in several countries, while, from the environmental perspective, this seems rather questionable. As a final example of promising policies in this area, the EU could also take advantage of the different rates and division of products in the VAT schemes. For instance, it is currently envisaged that energy products be subject to the lower rather than the normal VAT rate, despite the latter tariff would be preferable from an environmental perspective.

With respect to other environmental taxes or tradeable permits, interesting opportunities exist for environmental regulation through the tax system as well, such as taxes on the use of water or on highly polluting materials like heavy metals or chlorine. As recent surveys on the use of economic instruments show, governments in different countries of the EU are now paying much more attention to these opportunities. This can be seen as a promising trend as far as these taxes are aimed at internalizing existing externalities. Although such developments are contrary to the general trend of a shift from specific to general consumption taxes, they fit rather well in the trend to tax transactions more heavily.

Given the preceding analysis, it is not easy to assess the prospects of such a tax reform from a more general welfare point of view. Much depends on how much revenue is raised and how exactly the revenue is recycled. As long as environmental goals determine environmental tax reform proposals, thus yielding the first dividend, the revenue raised by such taxes will not be substantial in most cases. This is not surprising because the corrective tax principle of environmental tax reform is at least partly opposed to other tax principles such as tax neutrality. In particular, if the environmental tax does work for the environment, it cannot meet the principle of tax neutrality at the same time and vice versa. If the tax is raised on inelastic products, the Ramsey tax rule for tax neutrality is met, but whether the environment would benefit at all from such a tax could be seriously questioned. Hence, the government faces a trade-offs between corrective and neutral taxation. Energy taxation seems to be the only exception to this rule, due to the large amounts used as inputs in modern economies and the importance of the externalities (see also Chapter 3). Thus it is not surprising that energy taxation is popular for those seeking double dividends, for opportunities for a balanced budget tax reform only exist if enough revenue is raised.

Although principles other than the corrective tax principle can be considered as of secondary importance in the present context, they are crucial in the current debate on environmental tax reform. Because environmental taxes raise revenues, at least at the margin, and the tax liable agents do not know to what extent they will be compensated afterwards, implementation of such taxes often evokes pleas for countermeasures aimed at alleviating unfavorable consequences. If a particular tax system is made more corrective or regulatory, conflicts with the ability-to-pay principle arise more easily. To illustrate, poor and elderly people who live in old houses may suffer more than proportionally from a rise in the energy tax, while compensation can only alleviate part of this extra burden. Accordingly, environmental tax reform may hurt people with the lowest incomes thereby stimulating the call for countermeasures. Violation of the ability-to-pay principle may thus cause important political obstacles. As discussed in Chapter 3 the use of grandfather rules in tax reform, for instance by using tax exemptions, might also be used to abolish the obstacle of changes in the income distribution (Zodrow, 1992).

Yet environmental tax reform would offer opportunities for overall welfare improvement if these environmental taxes are envisaged not separately but as a package instead. In that case, the political authorities might take advantage of the fact that different taxes hurt agents differently. This provides opportunities for spreading the tax burden equally across different industries and households, thus creating more room for a Pareto-improving tax reform (Feldstein, 1976; Vollebergh and De Vries, 1992). This argument contains an important lesson for environmental tax reform and its policy implementation: presented as a package, acceptability of tax burdens could be negotiated through other mixtures of environmental taxes, while keeping the incentive as a whole intact. Moreover, compensation through reductions in other taxes might also be easier in the case of a package than if each particular tax were to be negotiated separately. Thus environmental tax reform might spread the burden more evenly across tax payers if a number of taxes are considered at the same time.

A final set of comments relates to the issue of capital mobility which might thwart any serious attempt to internalize externalities through an environmental tax reform program in the EU. Here, the effects depend on the complicated relationships between exposed sectors, the degree of competition and which environmental issue is taxed. As, for instance, a sector is exposed to EU-wide competition but not to competition with industries outside the EU, opportunities for EU policy are more promising. At the EU level this is true for rather different activities compared to level of individual Member States. For instance, intra-community flights can hardly compete with intercontinental flights, thus providing opportunities for an EU kerosine tax. Similar reasoning applies to agricultural products in relation to nutrient losses (Chapter 2), transport of persons and goods (Chapters 6 and 7), and at least major parts of the waste market (Chapter 5). Furthermore, as discussed extensively in Chapter 3, if the EU or the Member States want to act for the benefit of mankind as a whole (e.g. curbing CO_2 emissions), many opportunities exist to exempt internationally exposed sectors as is now already the case in several proposals.

Finally, adjustment costs due to an environmental tax reform could be minimized if certain implementation rules were followed. First, implementation should be announced beforehand. This enables the polluters to prepare for the new tax regime. Second, the tax should be implemented gradually so that adjustment could take place smoothly. Third, the package as a whole should be as transparent as possible, thus minimizing the risk of wage-price spirals. This, of course, assumes commitment from all parties to the agreement which is supposed to benefit all (due to the environmental benefits involved). Fourth, any industry facing international competition should be indulged in some way, while further political investment is made in global coordination of policies aimed at internalizing global environmental problems. Fifth, any environmental tax reform should be counted against other forms of regulation including non-regulation and their associated costs and benefits. With these lessons in mind, obstacles standing in the way of an environmental tax reform can be removed and opportunities be utilized. In that case, prospects for a welfare-improving European environmental tax reform do exist.

Notes

[1] We are grateful to Lans Bovenberg, Sijbren Cnossen, Frank Dietz and especially Jan de Vries for comments and Elbert Dijkgraaf for research assistance.

[2] This is sometimes loosely called tax neutrality (COM (92), 226). As will be discussed in the third section 'tax neutrality' refers to the tax policy principle of non-distortionary taxation, while the balanced budget assumption usually refers to cases in which the overall tax burden remains constant.

[3] Energy taxes as a percentage of total tax revenues are based on GREEN and provided by Hoeller and Wallin (1991, p. 28). For current tax profiles compare the next section.

[4] Table 1.1 in Chapter 1 provides an overview of those targets according to the EU.

[5] As noted by Messere (1993, p. 56ff), further problems arise because the information in these surveys is self-reported by countries, although with systematic verification by the OECD and IMF staff. Because countries differ with respect to their statistical reliability, the data might be biased although perhaps in an erratic way. Second, international comparisons are usually heavily influenced by the chosen numeraire, usually some measure relative to Gross Domestic Product (GDP). By using unweighted tax shares we avoided this problem, although this resulted in the exclusion of differences in GDP and the overall tax burden between countries.

[6] See e.g. Musgrave and Musgrave (1989, pp. 211-314) and Atkinson and Stiglitz (1980) for an overview of the most important tax policy principles and their application.

[7] Even transfers can be 'productive' in the sense that they may contribute to social welfare if some value for redistribution (i.e. equity) exists.

[8] Whether lump-sum taxation is possible in practice is not easy to answer. The 'head tax', which is levied as an equal amount of money per capita, is only a good example in specific circumstances. Lump-sum taxation should be considered a theoretical construct which is useful only as a counterfactual (Kay, 1990).

[9] Welfare analysis of environmental taxation usually does not take account of this fact. Polinsky and Shavell (1982) and Vollebergh (1994) are some exceptions.

[10] Sandmo (1975) is an exception.

[11] See Bovenberg (1995a) and Goulder (1994) for summaries of this literature.

[12] According to Schumpeterian economics, the tax rise on pollution acts as a signal to producers and consumers to invest in less polluting technological trajectories.

[13] However, higher income groups spend a larger budget share on gasoline, which may compensate for this effect if gasoline is also taxed. A similar argument holds for kerosine, as rich people fly more than poor people.

[14] Equity is exactly the reason why governments do not adopt lump-sum taxes but use distortionary taxes instead to raise public revenues. Environmental taxes may have a lump-sum element in taxing away transfer incomes. Note that environmental taxes may also be an indirect way to tax away the incomes from the informal economy (see Bovenberg and Van der Ploeg, 1995). A redistribution of incomes from the informal to the formal economy can be desirable from both a political and an economic point of view.

[15] Sometimes, other types of tax shifting are suggested which can be favorable for employment, e.g. by targeting the revenue recycling on low-skilled labor. Nevertheless, the effects of this targeting are also highly uncertain for the economy because they may change the income distribution

considerably (Bovenberg, 1995b).

[16] Compare Smith (1994) for an analysis of subsidiarity in relation to environmental taxation.

[17] Adverse effects for the environment itself are only to be expected if the environmental externality exceeds the jurisdiction of the EU as a whole - with benefits obtained largely by those outside the EU - and no countermeasures are taken to offset these effects.

[18] This might be brought under the assessment of support to industries at odds with the intention of the European Union Act which explicitly aims at equal competitive terms for industries across the EU.

References

Atkinson, A.B. and J.E. Stiglitz (1980), *Lectures on Public Economics,* McGraw Hill, New York.

Baumol, W.J. and W. Oates (1988), *The Theory of Environmental Policy,* Cambridge University Press, Cambridge.

Bovenberg, A.L. (1995a), 'Environmental Policy, Distortionary Labor Taxation and Employment: Pollution Taxes and the Double Dividend.' In: C. Carraro (ed.), *Frontiers of Environmental Economics,* forthcoming.

Bovenberg, A.L. (1995b), 'Environmental Taxation and Employment', *De Economist,* forthcoming.

Bovenberg, A.L. and R.A. de Mooij (1994a), 'Environmental Levies and Distortionary Taxation', *American Economic Review,* Vol. 94, Nr. 4, pp. 1085-1089.

Bovenberg, A.L. and R.A. de Mooij (1994b), 'Environmental Policy in a Small Open Economy with Distortionary Labor Taxes: A General Equilibrium Analysis'. In: E.C. van Ierland (ed.), *International Environmental Economics,* Elsevier Science Publishers, Amsterdam.

Bovenberg, A.L. and F. van der Ploeg (1994), 'Environmental Policy, Public Finance and the Labor Market in a Second-best World', *Journal of Public Economics,* Vol. 55, pp. 349-390.

Bovenberg, A.L. and F. van der Ploeg (1995), *Tax Reform, Structural Unemployment and the Environment,* mimeo, Tilburg University, Tilburg.

Burniaux, J-M. et.al. (1991), *The Costs of Policies to Reduce Global Emissions of CO_2: Initial Simulations with GREEN,* OECD Working Papers, No. 103, OECD, Paris.

CPB (Central Planning Bureau of the Netherlands) (1992), *Economische gevolgen op lange termijn van heffingen op energie* (Long-term Economic Consequences of Energy Taxes), Working Paper 43, CPB, The Hague.

European Economy (1992), *The Economics of Limiting CO_2 Emissions,* Special Edition Nr.1.

Feldstein, M. (1976), 'On the Theory of Tax Reform', *Journal of Public Economics,* Vol. 6, pp. 77-101.

Goulder, L.H. (1994), *Environmental Taxes and the Double-Dividend: A Readers Guide,* Paper presented at the 50th IIPF congress, Harvard, Cambridge MA, August 1994.

Goulder, L.H. (1995), 'Effects of Carbon Taxes in an Economy with Prior Tax Distortions: An Intertemporal General Equilibrium Analysis for the US', *Journal of Environmental Economics and Management,* forthcoming.

Harberger, A.C. (1962), 'The Incidence of the Corporation Income Tax', *Journal of Political Economy,* Vol. 70, pp. 215-240.

Hoel, M. (1991), 'Global Environmental Problems: The Effects of Unilateral Actions Taken by One Country', *Journal of Environmental Economics and Management,* Vol. 20, pp. 55-71.

Hoeller, P. and M. Wallin (1991), *Energy Prices, Taxes and Carbon Dioxide Emissions,* OECD Working Papers, Nr. 106, OECD, Paris.

IEA/OECD (1994), *Taxing Energy: Why and How?,* OECD, Paris.

Kay J. (1990), 'Tax Policy: A Survey', *Economic Journal,* Vol. 100, pp. 18-75.

Koopman, G.J. (1994), *Differential Treatment of Sectors and Energy Products in the Design of a CO_2/Energy Tax: Consequences for Employment, Economic Welfare and CO_2 Emissions,* paper presented at the workshop 'Environmental Taxation and Revenue Recycling', Eni Enrico Mattei Foundation, 16-17 December 1994, Milan.

Messere, K. (1993), *Tax Policy in OECD Countries,* IBFD Publications, Amsterdam.

Musgrave, R. and P.B. Musgrave (1989), *Public Finance in Theory and Practice,* fifth edition, Mc Graw Hill, New York.

Oates, W.E. (1991), *Pollution Charges as a Source of Public Revenues,* Discussion Paper QE92-05, Resources for the Future, Washington DC.

Oates, W.E. and R.M. Schwab (1988), 'Economic Competition among Jurisdictions: Efficiency Enhancing or Distortionary Inducing?', *Journal of Public Economics,* Vol. 35, pp. 333-354.

OECD (1989), *Economic Instruments for Environmental Protection,* OECD, Paris.

OECD (1993), *Environmental Taxes in OECD Countries: A Survey,* OECD Environment Monographs, Nr. 71, Paris.

Parry, I.H. (1994), *Pollution Taxes and Revenue Recycling,* mimeo, US Department of Agriculture, Washington DC.

Pearce, D.W. (1991), 'The Role of Carbon Taxes in Adjusting to Global Warming', *Economic Journal,* Vol. 101, pp. 938-948.

Polinsky, A.M. and S. Shavell (1982), 'Pigouvian Taxation with Administrative Costs', *Journal of Public Economics,* Vol. 19, pp. 385-394.

Sandmo, A. (1975), 'Optimal Taxation in the Presence of Externalities', *Swedish Journal of Economics,* Vol. 77, pp. 86-98.

Schöb, R. (1994), *Evaluating Tax Reforms in the Presence of Externalities,* paper presented at the 50th IIPF congress, Harvard, Cambridge MA, August 1994.

Smith, S. (1992), 'Taxation and the Environment: A Survey', *Fiscal Studies,* Vol. 13, Nr. 4, 21-57.

Smith, S. (1994), *Federal Issues in Environmental Taxation,* paper presented at the 50th IIPF congress, Harvard, Cambridge MA, August 1994.

UPI (Umwelt- und Prognose-Institut) (1988), *Okosteuern als Marktwirtschafliches Instruments im Umweltschutz*, Heidelberg.

Vollebergh, H.R.J. (1994), *Environmental Taxes and Transaction Costs*, Tinbergen Institute Discussion Paper 94-96, Erasmus University, Rotterdam.

Vollebergh, H.R.J. (1995), 'Transaction Costs and European Carbon Tax Design'. In: M. Faure, J. Vervaele, and A. Weale (eds.), *Environmental Standards in the European Union in an Interdisciplinary Framework*, MAKLU et.al., Antwerpen, pp. 135-154.

Vollebergh, H.R.J. and J. de Vries (1992), 'Financiële gevolgen van een regulerend milieuheffingen-pakket (Financial Effects of a Package of Regulating Environmental Taxes)', *Tijdschrift voor Politieke Ekonomie*, Vol. 15, Nr. 2, pp. 9-30.

Wit, G. de (1994), *Werkgelegenheidseffecten van een belastingverschuiving van arbeid naar milieu (Employment Effects of a Shift from Labor towards Energy)*, Centre for Energy Conservation, Delft.

Zodrow, G.R. (1992), 'Grandfather Rules and the Theory of Optimal Tax Reform', *Journal of Public Economics*, Vol. 49, pp. 163-190.

9 A Comment on Political Feasibility

Alex de Savornin Lohman

On the Political Feasibility of Policy Proposals

The present volume is a collection of policy proposals (to be referred to as PPs) that have been designed by members of a working group on environment and economy from the National Environmental Forum (Landelijk Milieu Overleg - LMO), a coordinating organization of the Dutch environmental movement. The goal of this paper is to assess the *political feasibility* of the PPs presented.

The PPs are all economic instruments, or rather economic *incentives*, as they rely on financial incentives to change the behavior of polluters. Among the PPs are three incentive charges (the solvent tax, the surplus levy on nutrient emissions and the discussion of fiscal instruments for transport policy), two tradeable permit systems (tradeable landfill certificates and tradeable transit rights in the EU-Austrian treaty on transit traffic) and hybrid incentive mechanisms to abate carbon emissions.

The PPs are in different areas of environmental policy. The degree to which the PPs have been worked out in terms of actual design (charge rates, number of permits) and implementational detail (monitoring and enforcement) differs considerably. Some PPs have extensive discussions of implementational aspects (tradeable transit ecopoints and the solvent tax), others devote some attention to it without going into detail (surplus levy on nutrient emissions, carbon incentive mechanisms and fiscal instruments for transport policy) and one deals only briefly with implementation aspects (tradeable landfill certificates). None of the PPs is worked out in such detail, that it could figure as the basis for a bill of law. This may not have been the objective of the authors, but a lack of detail does complicate the task of the commentator, as with regard to political feasibility the devil is often in the detail. Nevertheless, an attempt will be made, taking the essence of the PPs as the object of discussion.

An initial observation is that *in their present form* the political status of the PPs is non-existent, with one exception.[1] The PPs are as yet only texts, that have not been fed into the policy circuit. Kelman (1981), in an empirical study of the political fate of proposals for economic incentives in US environmental policy, concluded that a *consciously designed political strategy* is required for PPs to be seriously considered by the policy circuit. The strategy must be explicitly directed towards carefully selected chosen policy actors. To put it in marketing terms: the policy product must be sold to the relevant target group.

What would such a strategy look like? Meltsner (1972) offers a framework for designing a political strategy, starting from 'a mapping of policy actors, their beliefs and motivations, resources, and the sites of their interactions'. A 'political feasibility map' drawn up along these lines would provide the building blocks for a strategy. The strategy would have to be contingent on the specifics of the political system in question, the characteristics of the policy arena surrounding the environmental policy issue at hand as well as its wider ramifications.

PPs have to be fed into the policy circuit at the right time, at the right place and to the right actor. Policy making is not an intellectual process in which a single 'policy maker' works out *substantive* solutions to preconceived problems, like a Platonic philosopher-king. First and foremost, policy making consists of a set of *procedural* rules defining who is allowed access to the policy making process, the distribution of competences among policy actors, the timing of the steps in the policy making chain and the fora for interaction of policy actors. Policy making is a social process, in which flesh-and-blood policy actors interact, each with their own set of perceptions, visions, interests and resources. This has to be taken into account in designing a strategy that is geared to the specific characteristics of the policy issue at hand.

The first step in a political strategy is the presentation of the PP to policy actors, with the objective of persuading the actor concerned of the merits of the proposal. The actor must be convinced that the PP 'has something in it for him'. As Majone has argued, persuasion is a different matter from scientific exposition (Majone, 1991). One important difference is, that to be persuasive a text must take the frame of reference of the actor concerned as its point of departure.

Now what is the frame of reference shared by policy actors? Is it 'the environmental problem at stake'? The answer is *no*, as different policy actors have different perceptions of 'the problem at stake'. Some think it exists, others deny that. Some derive a benefit from policy actions, others are hurt by them. Some think the issue at stake has wider ramifications, others consider it in a restricted framework. Some think the issue is urgent, others think an active policy can wait a while.

But there is one thing policy actors do have in common, and that is *the existence of actual policies*. Actual policies - also if they happen to be particularly ineffective or symbolic - are existing things, and it is around the progressive change of actual policies that networks of policy actors evolve. Policy actors define their positions and act in relationship to actual policies, not only because that is their short term interest, but also for the reason that actual policies are the common ground they have.

A PP that wants to be taken seriously by relevant policy actors, will first of all have to clarify its position vis a vis existing policies. Therefore the PP must give answers, in a manner that is clear cut to the policy actor concerned, to questions such as: in what sense does the PP differ from actual policies; is it designed as a replacement of actual policies, as an add-on or as an amendment? What defect of actual policies does the PP intend to remedy: effectiveness, efficiency, problems of monitoring and control, distributional equity or something else? Is its (implicit) environmental target equal to, stronger or weaker than the (implicit) target of actual policies? Can existing institutions implement the PP, or are new institutions necessary? If so, what type of institutions and can a meaningful political trajectory be indicated towards the creation of such institutions? Who bears the costs and who derives the benefits of the PP, in comparison to the existing distribution of costs and benefits?

Not all these questions can be answered for the PPs presented in this volume, if only because the proposals have not been worked out to such a level of detail that answers can be given. The remainder of this paper is an attempt to address *some* of the issues, consecutively for each PP. One PP is not discussed here, namely the hybrid incentive mechanisms for CO_2 abatement, as this proposal already devotes considerable attention to political feasibility. The only issue that remains is the implementability of the PP, but this cannot be treated satisfactorily here, for reasons of space.

The PPs will be discussed in the following format: for each PP the first step will be to restate it in terms of its relationship to actual policies. The next step is to devise an answer, if possible, to the questions raised above. If the PP concerned leaves certain critical questions unanswered, this is indicated: how can the PP be strengthened, by adding additional detail and/or supporting evidence, to be as convincing as possible? Finally, the outlines of a strategy are indicated for making the PP go as far as possible in the policy circuit.

The Policy Proposals

Tradeable Landfill Certificates

The main part of Chapter 5 consists of a critical appraisal of EU policies for household and industrial waste. The issue at stake is whether existing policies are in conformity with the principles for waste policy that the EU itself has formulated. Existing EU policies consist of a set of directives that have been promulgated or are in preparation. The directives refer to environmental requirements for waste processing, and the tariffs to be paid. The EU policy principles are twofold: the first principle is the 'priority ladder', with waste prevention as the best option, next recycling, followed by incineration, and finally landfilling as the least preferred option. The second principle is the so-called 'proximity principle', stating that wastes produced must be processed as near as possible to their origin.

The chapter goes on to demonstrate that existing EU policies are insufficient to attain agreement with the policy principles. A major obstacle is the conflict between the proximity principle and the EU Internal Market, allowing wastes to be transported across national borders. Any positive environmental effects of the waste directives are cancelled out by additional waste transports, that make wastes flow to the lowest point in the system: landfills in countries with low tariffs.

Adding an incentive charge on waste to existing directives would not be sufficient to attain conformity to the policy principles. Moreover, the charge would impose a significant tax burden on waste producers, even larger than the costs of waste removal and processing themselves. As existing policies are insufficient, and an incentive charge is also insufficient and economically too painful, the PP proposes an EU-wide cap on landfill capacity.

No doubt, this is a far-reaching proposal. In relation to existing policies the PP implies a considerable sharpening of the policy objective, going beyond policies that have so far been formulated by Member States themselves. The pill is sweetened for waste producers by making the right to deposit wastes on landfills tradeable, and by grandfathering these depositing rights to landfill owners and/or 'existing waste producers'.

The economic impact of the PP on waste producers has not been worked out. Although the permits are to be grandfathered, the economic impacts will still be substantial due to the impact of the cap on landfilling capacity. In addition, the PP raises a number of questions as to implementability (can a stop on landfilling capacity be enforced?) and institutional requirements, because a supranational 'EU waste management agency' would be required, strong enough to impose and implement the policy in all Member States.

Assessing the economic impacts, the implementational problems and institutional requirements are issues the PP should address first, in order to be considered as a full-fledged policy proposal. In its present form the PP leaves these important issues unanswered. Nevertheless, in its present form the PP could be useful for another purpose. It exposes in a numerical fashion the contradictions inherent in the EU proximity principle and the creation of an Internal Market. As such, it could be valuable as ammunition to policy actors - such as members of the European Parliament - that are willing to take on the case of excluding waste from the open market principle.

Tradeability of Ecopoints in the EU-Austrian Transit Agreement

The PP is an amendment of the agreement on transit traffic that has been concluded between the EU and Austria, as a part of the treaty on Austria's conditions of entry into the EU. One of the elements of the transit agreement is a system of 'ecopoints', limiting both NO_x emissions from transit traffic and the number of trips. The PP proposes to make the ecopoints tradeable within the EU.

The PP is not far removed from existing policies. It takes over the environmental objective that is implicit in the transit agreement, that is the ceilings imposed on NO_x emissions from freight traffic and the number of trips. Moreover, no substantial change is required in implementation mechanisms and institutions. A new institution would be required to manage the market for ecopoints, but as the institution responsible would be little more than a brokerage house this requirement does not look insurmountable.

Experience shows, that a tradeable permit system can be politically acceptable if the permits are grandfathered.[2] Surely the details of the formula for grandfathering, i.e. the initial distribution of ecopoints between (and within!) Member States would lead to a political battle now that the ecopoints get a monetary value. The prospect of inter Member State conflict on the formula for initial distribution of ecopoints may make EU officials reluctant to adopt the proposal.

As concerns political feasibility, the proximity of the PP to existing policies is a strong point. Compared to the ecopoint system contained in the existing transit agreement, there would be economic benefits to transport companies. Will these benefits actually occur? The essential question is whether transport companies will actually put ecopoints on the market. Will companies speed up fleet renewal beyond the pace prescribed by the progressive overall reduction of ecopoints, in expectation of finding a buyer and of getting a sufficient price for the ecopoints thus earned? In other words: will the market be sufficiently developed for the ecopoints to have liquidity in the Keynesian sense?

Of course, firms going banktrupt will put ecopoints on the market, but such ecopoints would also be available, at zero price, under the present system of non-tradeable points. New firms, that do not benefit from the grandfathering, will have to buy themselves in - a competitive disadvantage that may stifle the economic and environmental dynamism of the branch.

Another political complexity is the uncertain future of the ecopoint system. Will it be continued after the six years' reappraisal? If not, a reform introducing tradeability will only be valid for a few years and this may not be worth the political energy to be spent. In principle, Austria has a veto position on the date of renewal. But once it is an EU member, it will come under stronger pressure, politically and legally, to abolish any remaining border controls including the ecopoint system.

Concerning strategy, the PP could best be presented to those policy actors who can expect to benefit from it. Primary target groups are associations of transport companies, and government officials from those Member States who will be most affected by the progressively falling number of ecopoints available, such as Germany, Greece and Italy. To be convincing, the PP should offer more certainty on the economic benefits to be expected, for instance by including evidence from existing tradeable permit systems.

For the sake of equity and of maintaining industrial dynamism the PP should have a provision for newcomers on the transport market, such as the US Acid Rain Control

Allowance System. For instance, the points that come on the market from firms going banktrupt could be put in a reserve to be made available for new firms.

The Solvent Tax

This PP has been worked out in the context of an assignment by the Environment DG of the EU, with the objective of covering VOC emissions from smaller sources, that are not affected by the existing draft Directive on large sources. The large sources Directive is a combination of direct regulation and voluntary agreements, an approach that would clearly be infeasible for smaller sources.

The PP contains an extensive discussion of fiscal implementation aspects, especially the question whether a tax on kilograms of produced and imported solvents or a tax on products containing solvents would be most practicable. The PP comes out in favor of a tax on solvents, with a rate of 1 ECU per kg. The PP estimates that this rate should be sufficient for a limitation of emissions of ca. 30%, which is the official EU policy objective. The tax would be refunded for exports, and be levied on imports according to administrative estimates of solvent content.

The problem of a uniform tax rate is that the environmental impacts of solvent use are not uniform. Tax exemptions are envisaged for use as feedstock and for recycled solvents, and refunds are to be given in case of 'proven' emission reduction efforts. Refunds for types of usage 'for which no alternative is feasible' are rejected, as this would involve the authorities in precisely the type of discussion that environmental taxes are meant to avoid. Presumably, these refunds have been suggested to the EU environmental officials by representants of industries producing and using solvents. These industries are by nature opposed to the tax, precisely because it is a generic policy instrument that is indiscriminately (no bargaining allowed!) valid for all usage, and as it involves a transfer to the public budget, which is absent in direct regulation and voluntary agreements. The transfer could be diminished, and thus the resistance by affected industries, by using the tax proceeds to provide subsidies for reduction of solvent emissions.

A political obstacle to the tax is that larger industries do have a point in arguing that they are 'hit by two sticks', as the tax would come on top of the large sources Directive. The PP argues that large industries should not be exempted from the tax, but a better solution would be to drop the large sources Directive entirely. However, this option will surely meet resistance with the EU environment directorate. Environmental officials, as most government officials, tend to be protective of existing policies.

Another source of resistance comes from Member States that have a low priority for VOC reduction policies. The PP suggests that these states could be bribed by allocations out of the solvent tax' revenues.[3] For this purpose (part of) the revenue would have to be channeled to the EU. Whether this is politically and administratively feasible is an open question.

Conceivably, Member States also having their own policies in place (predominantly voluntary agreements) may take the view that they prefer to await the results of their ongoing programs. So the PP could be strengthened by a demonstration that existing Member State policies are in fact insufficient.

The solvent tax comes up against the 'paradox of effectiveness', that is well known in environmental policy. It is a relatively simple, generic policy instrument, that definitely has a bite, and precisely for these reasons it will conjure up a maximum of resistance. The resistance may be outright, or come in the guise of 'technical' criticism, such as on compatibility with GATT provisions, tax principles, etc. Still, it is difficult to see how the problem of VOC emissions from non-point sources could be addressed by another policy instrument. Therefore it is definitely worthwhile to address the remaining technical complexities, and to give thought to a political strategy to make the PP go as far as possible in the policy circuit.

A strategy would have to make the case for EU action as clear as possible, in view of the present popularity of the subsidiarity principle. Concrete options should be given to channel the tax revenue back to polluters by means of a combined tax-subsidy. The problem of the 'double burden' for large sources must be addressed, although it is difficult to see how this could be done without relinquishing the existing large sources Directive. Finally, the PP should be clearer as regards the expected environmental effects (the elasticity of the tax).

The Surplus Levy on Nutrient Emissions from Agriculture

Present EU policies for reducing nutrient emissions from agriculture are formulated in terms of uniform emission standards specifying allowable application rates of nitrogen per hectare. The choice of policy instruments to attain these emissions standards is left to EU Member States.[4] Present Member State policies rely on a mixture of direct regulation, mostly prescribing specific technologies to be applied, manure storage and/or manure processing. Financing of these policies is either by the agricultural sector itself or by subsidies.

The problem with present direct regulation policies is that they leave no scope for choice of technology at farm level. The technologies prescribed are often of an end-of-pipe character, and there is little incentive to develop new technologies for emission abatement. In addition, there are considerable problems in monitoring and enforcing the complex array of regulations.

All in all, this seems to be a classic case for which economic incentives could be fruitful in environmental as well as economic terms. The economic incentive that is mostly discussed in the literature is an input levy on fertilizer. The PP criticizes the fertilizer levy, as it is on the input and not on the emission of nutrients, containing no incentive to reduce nutrient losses at farm level. Moreover, a uniform levy does not take into account local differences in the carrying capacity of nature. Also, a product levy could be considered as a barrier to trade and acts as an implicit premium on illegal parallel imports of fertilizer.

The PP is an emission charge, to be levied on the so-called nutrient surplus, i.e. the difference between actual nutrient emissions and acceptable emission standards that are differentiated for different types of land.[5] The nutrient surplus is equal to actual emissions as measured by a 'nutrient bookkeeping system', to be kept at farm level, and the emission standards that are valid for the specific land type in question. Farmers can apply for exemptions to the standards, if they can prove that on their land a higher standard is allowable without undue damage to the environment.

Compared to present direct regulation policies, the PP shifts a good deal of agency costs for monitoring and enforcement to the farmer: the costs of keeping the nutrient bookkeeping system, and the costs of applying for exemptions. This is in accordance with the polluter pays principle and with other permit policies, in which the costs of self-reporting are mostly borne by sources themselves. The agricultural sector will not like this feature of the PP, but it is simply a matter of equal treatment compared to other industrial sectors.

The proposed rate for the surplus levy has been derived from various studies on elasticities, but it remains uncertain to what extent the proposed rate will be successful in actually reducing emissions.[6] Moreover, the PP accepts that for some types of agriculture (with a high value added per unit of emissions) the standards will be exceeded. An alternative could be to impose the levy on (nearly) all emissions, while rebating the revenue to the sector concerned, e.g. on the basis of output.

The political feasibility of the PP will depend strongly on the rate proposed, and the spending of the revenue. There will be strong resistance from the agricultural sector anyhow, as the sector seems to prefer present direct regulation policies on which it expects to have greater influence in the implementation phase. Moreover, the levy will worsen the competitive position of the northern Member States that has already been negatively affected by the 1992 reform of the common agricultural policy.

On the other hand, it is becoming increasingly clear that (a) something serious has to be done about nutrient emissions (cf. also the EU commitment to reduce nutrient emissions to the North Sea) and (b) present policies are rather expensive for farmers (witness the failure of the manure processing plants in the Netherlands). Therefore it is logical that policy discussions tend more and more towards economic incentives, where indeed a choice between an input levy and a levy on emissions is to be faced. In this discussion, an emission charge - whether on actual emissions or on the nutrient surplus as in the PP - has certain advantages, as has been indicated above. Critical points are whether the nutrient bookkeeping system can really be made to work, how the rate of the levy should be determined on the basis of elasticity studies and whether a method can be found to plough the revenue back to the sectors concerned.

Fiscal Instruments for Transport Policy

The discussion of fiscal instruments for EU transport policy is of interest as it gives a clear picture of the complexity of the issue. There are many environmental problems related to transport, and these problems are interconnected in complex manners. Moreover, there are non-environmental external costs of transport: safety, land use and congestion. Finally, the transport system is highly constitutive for the structure of EU economies, determining industrial location patterns and citizen lifestyles.

Because of the inherent complexity of the transport issue, the standard decision making approach of 'first defining the problem' may not be the appropriate course. Rather, finding a 'satisficing' policy instrument that has an impact on various interconnected problems (without completely solving either of them) will make the most of the scarce time that is available on the EU political agenda. One should keep in mind, that resources for decision making are finite. It is always easier to jump on a moving train: adding environmental policies to existing processes of economic decision making, such as fiscal harmonization and product standardization, enables EU environmental officials to use the momentum of the creation of internal market for their own purposes.

So far EU transport-environment policy has been most successful in emission standards for vehicles. An interesting aspect of EU policies in this field is that a way has been found to accommodate the desires of Member States wishing to go further than the standards agreed upon in community decision making: tax differentiations on the basis of future (tighter) standards. Successful precedents should always be used, and extended.

In the EU fiscal harmonization of car petrol excises, a pragmatic coalition could be forged of environmentally minded northern states with southern Member States, which historically have high excises. Surely, bringing the environment in on the fiscal harmonization issue will require political and administrative groundwork to improve the position of the environment DG within the EU bureaucracy. Presently, the environment DG is notoriously weak, being understaffed as well. A lot needs to be done here on institution building.

Scattered throughout the text on fiscal instruments are references to 'resistances' of a socio-political nature. Indeed, any policy proposal in this area will come up against strong counterforces. The enormous growth of the transport sector, made possible by an array of implicit subsidies, has created vested interests in road haulage and has changed industrial location patterns and citizen lifestyles in a manner that is not easily reversed. We are now saddled with a transport infrastructure that cannot be changed overnight, and has served as the groundwork for additional irreversible investments in an economic superstructure with high mobility demands. Moreover, the unhampered growth of mobility has infected EU citizens with the notion of a 'right to mobility', a notion that is reinforced time and again by messages in the media associating mobility with freedom. Reversing this trend in Europe will command more than the design of a set of policy instruments. Without a broader societal debate on environmental limits to transport, policy proposals with a real bite will be smothered in road blocks by truck drivers and voter walkouts. The debate will have to be on the principles of an ecological modernization of the European transport infrastructure, on a conscious industrial policy to promote the technologies required and, last but not least, on an environmental regime for the cuckoo in the transport nest: air traffic.

Concluding Remarks: Action or Reaction?

The purpose of this chapter is not to pass a final judgment on the political feasibility 'yes' or 'no'. That would be an overly static conception of political feasibility. The essential thing is to get communication with crucial policy actors started, and it should be realized then that any communicative process worth its name contains the germs of redefinition of positions. In communicating a PP to policy actors the PP will be reframed, as policy actors will come up with new constraints that have not been taken into account: practical constraints, distributional constraints and institutional constraints. If the communicative process with policy actors is working at all, the PP will be reframed successively with an endpoint that is impossible to predict. In such a process, environmentalist organizations who actively put forward concrete policy proposals make themselves vulnerable to criticism on the technical aspects of policy making.

Criticizing existing environmental policies in a reactive manner is one thing, but developing working alternatives is another. It is no mean thing to elaborate a real policy proposal that is sufficiently worked out to function as the basis for a bill of law. Normally the design of concrete policies, including all the required administrative detail, is the task of state agencies, that, for one thing, are equipped with the resources to do the job. So, a strategic question would have to be answered first: Is the development of concrete policy proposals a fruitful strategy for environmentalist organizations?

The answer could be that environmentalist organizations should limit themselves to participating in policy circuits in a *reactive* manner only, trying to make the best of policies that are being shaped in the interplay between policy actors. Or perhaps the optimal strategy is to bypass official policy circuits and go for the traditional strategies of the environmental movement: societal mobilization and agenda formation.

But the anwer could also be that environmentalist organizations should (acquire and) devote more resources to the task of elaborating working policy proposals. This is the approach followed in the United States by organizations such as the Environmental Defense Fund, which played an active role in the shaping of the tradeable permits system for Acid Rain Control now functioning in the US. This is also the approach followed in the preceding chapters.

This contribution does not address these strategic questions. The point to be made here is only that the design and presentation of policy proposals is not a good thing in itself. It has to be the outcome of a strategic choice - one that has consequences in terms of time and resources and political status of the environmental movement.

Notes

[1] The exception is the PP for a solvent tax, that has been worked out as an assignment from the Environment DG of the European Commission. Therefore this PP is already inside the policy circuit.

[2] In all existing tradeable permit systems (mainly in the United States) permits are grandfathered, thus avoiding the transfer from polluters to the public budget that would be entailed by auctioning the permits.

[3] The PP also suggests an allocation from the Cohesion fund, but the Member States concerned will consider this a 'cigar out of their own pocket'.

[4] The control of phosphate emissions is also left to the discretion of Member States, as phosphate emissions do not constitute a serious environmental problem in all Member States.

[5] The standards proposed by the authors are somewhat different from the standards defined in the EU nitrate directive, but this is immaterial to the essence of the PP.

[6] Of course, the environmental effectiveness of direct regulations prescribing specific technologies is also uncertain as long as monitoring and enforcement are imperfect, which they will always be.

References

Kelman, S. (1981), *What Price Incentives?*, Auburn House, Boston Mass.

Majone, G. (1991), *Evidence, Argument and Persuasion in the Policy Process*, Yale University Press, New Haven.

Meltsner, A.J. (1972), 'Political Feasibility and Policy Analysis', *Public Administration Review*, November/December.

List of Contributors

Frank J. Dietz
Department of Public Administration
Erasmus University
P.O. Box 1738
3000 DR Rotterdam
The Netherlands

Pascale van Duyse
Institute for Applied Environmental Economics (TME)
Grote Marktstraat 24
2511 BJ The Hague
The Netherlands

Heddeke Heijnes
Institute for Applied Environmental Economics (TME)
Grote Marktstraat 24
2511 BJ The Hague
The Netherlands

Paul R. Koutstaal
Department of Economics
Faculty of Law
Groningen University
P.O. Box 716
9700 AS Groningen
The Netherlands

Joram Krozer
Institute for Applied Environmental Economics (TME)
Grote Marktstraat 24
2511 BJ The Hague
The Netherlands

Ruud A. de Mooij
Research Centre for Economic Policy (OCFEB) and Ministry of Economic Affairs
Erasmus University
P.O. Box 1738
3000 DR Rotterdam
The Netherlands

Xander Olsthoorn
Institute for Environmental Studies (IVM)
De Boelelaan 1115
1081 HV Amsterdam
The Netherlands

Frans Oosterhuis
Institute for Environmental Studies (IVM)
De Boelelaan 1115
1081 HV Amsterdam
The Netherlands

Alex de Savornin Lohman
Institute for Environmental Studies (IVM)
De Boelelaan 1115
1081 HV Amsterdam
The Netherlands

Herman R.J. Vollebergh
Research Centre for Economic Policy (OCFEB) and Faculty of Economics
Erasmus University
P.O. Box 1738
3000 DR Rotterdam
The Netherlands

Jan L. de Vries
Study Group on Environment and Economics
National Environmental Forum (LMO)
Donkerstraat 17
3511 KB Utrecht

Bert van Wee
National Institute of Public Health and Environmental Protection (RIVM)
P.O. Box 1
3720 BA Bilthoven
The Netherlands

Frans van der Woerd
Institute for Environmental Studies (IVM)
De Boelelaan 1115
1081 HV Amsterdam
The Netherlands

Subject Index

ENVIRONMENT & POLICY

1. Dutch Committee for Long-Term Environmental Policy: *The Environment: Towards a Sustainable Future.* 1994 ISBN 0-7923-2655-5; Pb 0-7923-2656-3
2. O. Kuik, P. Peters and N. Schrijver (eds.): *Joint Implementation to Curb Climate Change.* Legal and Economic Aspects. 1994 ISBN 0-7923-2825-6
3. C.J. Jepma (ed.): *The Feasibility of Joint Implementation.* 1995 ISBN 0-7923-3426-4
4. F.J. Dietz, H.R.J. Vollebergh and J.L. de Vries (eds.): *Environment, Incentives and the Common Market.* 1995 ISBN 0-7923-3602-X

KLUWER ACADEMIC PUBLISHERS – DORDRECHT / BOSTON / LONDON

MIX
Papier aus verantwortungsvollen Quellen
Paper from responsible sources
FSC® C105338

If you have any concerns about our products,
you can contact us on
ProductSafety@springernature.com

In case Publisher is established outside the EU,
the EU authorized representative is:
Springer Nature Customer Service Center GmbH
Europaplatz 3, 69115 Heidelberg, Germany

Printed by Libri Plureos GmbH
in Hamburg, Germany